Lecture Notes in Mathematics 1730

Springer
Berlin
Heidelberg
New York
Barcelona
Hong Kong
London
Milan
Paris
Singapore
Tokyo

Siegfried Graf Harald Luschgy

Foundations of Quantization for Probability Distributions

Springer

Authors

Siegfried Graf
Faculty for Mathematics and Computer Science
University of Passau
94030 Passau, Germany

E-mail: graf@fmi.uni-passau.de

Harald Luschgy
FB IV, Mathematics
University of Trier
54286 Trier, Germany

E-mail: luschgy@uni-trier.de

Cataloging-in-Publication Data applied for
Die Deutsche Bibliothek - CIP-Einheitsaufnahme

Graf, Siegfried:
Foundations of quantization for probability distributions / Siegfried Graf ; Harald Luschgy. - Berlin ; Heidelberg ; New York ; Barcelona ; Hong Kong ; London ; Milan ; Paris ; Singapore ; Tokyo : Springer, 2000
(Lecture notes in mathematics ; 1730)
ISBN 3-540-67394-6

Mathematics Subject Classification (2000): 60Exx, 62H30, 28A80, 90B05, 94A29

ISSN 0075-8434
ISBN 3-540-67394-6 Springer-Verlag Berlin Heidelberg New York

Springer-Verlag is a company in the BertelsmannSpringer publishing group.

Printed in Germany

Typesetting: Camera-ready TEX output by the author
Printed on acid-free paper SPIN: 10724973 41/3143/du 543210

Contents

List of Figures

List of Tables

Introduction

The term "quantization" in the title originates in the theory of signal processing. It was used by electrical engineers starting in the late 40's. In this context quantization means a process of discretising signals and should not be mistaken for the same term in quantum physics. As a mathematical topic quantization for probability distributions concerns the best approximation of a d-dimensional probability distribution P by a discrete probability with a given number n of supporting points or in other words, the best approximation of a d-dimensional random vector X with distribution P by a random vector Y with at most n values in its image. It turns out that for the error measures used in this book there is always a best approximation of the form $f(X)$, a "quantized version of X". The quantization problem can be rephrased as a partition problem of the underlying space which explains the term quantization.

Much of the early attention in the engineering and statistical literature was concentrated on the one-dimensional quantization problem. See Bennett (1948), Panter and Dite (1951), Lloyd's 1957 paper (published 1982), Dalenius (1950), and Cox (1957). Steinhaus (1956) was apparently the first who explicitly dealt with the problem and formulated it for general (3-dimensional) spaces. Since then quantization occurred in various scientific fields, for instance

- Information theory (signal compression): Shannon (1959), Gersho and Gray (1992)
- Cluster analysis (quantization of empirical measures), pattern recognition, speech recognition: Anderberg (1973), Bock (1974), Diday and Simon (1976), Tou and Gonzales (1974)
- Numerical integration: Pagès (1997)
- Stochastic processes (sampling design): Bucklew and Cambanis (1988), Benhenni and Cambanis (1996)
- Mathematical models in economics (optimal location of service centers): Bollobás (1972,1973)

The aim of the present book is to describe the mathematical theory underlying the different applications of quantization. The emphasis is on absolutely continuous as well as on singular (continuous) distributions on $\mathbb{R}^d$. In the nonsingular case we present a rigorous treatment of known results in quantization including various new aspects while the results for singular distributions seem to be completely new.

In more detail we consider the following problem.
We concentrate on norm-based error measures. Let $\| \ \|$ be a norm on $\mathbb{R}^d$ and $1 \leq r < \infty$. Define the Wasserstein-Kantorovitch L_r-metric ρ_r for probabilities P_1, P_2 by

$$\rho_r(P_1, P_2) = \inf\Big\{ \Big(\int \|x - y\|^r \, d\mu(x,y) \Big)^{1/r} : \mu \text{ probability on } \mathbb{R}^d \times \mathbb{R}^d \text{ with marginals } P_1 \text{ and } P_2 \Big\}$$

and the (minimal) quantization error of a given probability P and $n \in \mathbb{N}$ by

$$\begin{aligned} V_{n,r}(P) &= \inf \{ \rho_r(P, Q)^r : |\operatorname{supp}(Q)| \leq n \} \\ &= \inf \{ E \, \|X - f(X)\|^r : f\colon \mathbb{R}^d \to \mathbb{R}^d \text{ measurable}, |f(\mathbb{R}^d)| \leq n \}. \end{aligned}$$

For an optimal quantizing rule f, i.e. f attaining the inf, the domains of constancy provide P-almost surely a Voronoi partition of $\mathbb{R}^d$ with respect to its respective values. As a consequence the quantization problem is also equivalent to the n-centers problem, which requires finding a set of n elements α which minimizes the expression

$$E \min_{a \in \alpha} \|X - a\|^r$$

whose minimum value equals $V_{n,r}(P)$, that is

$$V_{n,r}(P) = \inf \{ E \min_{a \in \alpha} \|X - a\|^r : \alpha \subset \mathbb{R}^d, |\alpha| \leq n \}.$$

In Chapter I we present general properties of the quantization problem for a fixed number n of quantizing levels. We discuss existence of optimal quantizers, necessary conditions for optimality, and a sufficient conditon for uniqueness in the one-dimensional case. Under this uniqueness condition it is easy to find numerically optimal quantizers for P in the one-dimensional case. However, for dimensions $d \geq 2$ it is difficult even for small n to determine optimal quantizers. This is caused (among other things) by the fact that the minimization function $(a_1, \ldots, a_n) \mapsto E \min_{1 \leq i \leq n} \|X - a_i\|^r$ is (typically) nonconvex. As examples we consider uniform distributions on ball packings and spherical distributions. Furthermore, we prove stability properties which can be applied to the empirical analysis of the quantization problem.

Chapters II and III focuse on the asymptotic behaviour of the quantization problem when the number n of quantizing levels tends to infinity. Chapter II starts with the investigation of the order of convergence to zero for the sequence of quantization errors $(V_{n,r}(P))_{n \geq 1}$ for probabilities P on $\mathbb{R}^d$ whose absolutely continuous part does not vanish. It is shown that the limit of the sequence $(n^{\frac{r}{d}} V_{n,r}(P))_{n \geq 1}$ exists in $(0, \infty)$, provided a certain moment condition is satisfied. The limit

$$Q_r(P) = \lim_{n \to \infty} n^{r/d} V_{n,r}(P)$$

is called r-th quantization coefficient and can be expressed in terms of the r-th quantization coefficient

$$Q_r([0,1]^d) = \lim_{n\to\infty} n^{r/d} V_{n,r}\left(U\left([0,1]^d\right)\right)$$
$$= \inf_{n\geq 1} n^{r/d} V_{n,r}\left(U\left([0,1]^d\right)\right)$$

of the uniform distribution on the unit cube of $\mathbb{R}^d$ and the density of the absolutely continuous part of P with respect to Lebesgue measure. Quantization coefficients provide interesting parameters for probability distributions. They can be evaluated for univariate distributions and some of them also for multivariate distributions. Fundamental work is due to Fejes Tóth (1959) and Zador (1963). Next we define asymptotically optimal sequences of quantizers and sets of centers for nonsingular probability distributions P and investigate their properties. It is proved that the empirical measures corresponding to asymptotically optimal sets of centers converge weakly to a probability on $\mathbb{R}^d$ which is explicitly given using P. Furthermore, the asymptotic performance of certain classes of quantizers is compared to that of (asymptotically) optimal quantizers. In particular, we consider regular quantizers which are based on space-filling figures in $\mathbb{R}^d$, lattice quantizers, product quantizers, and random quantizers. The results provide bounds for the quantization coefficients $Q_r\left([0,1]^d\right)$.

All these considerations concern the case $1 \leq r < +\infty$ and arbitrary norms on $\mathbb{R}^d$. The rest of the chapter is devoted to the study of similar results for a geometric covering problem which corresponds to the case $r = \infty$. Here the quantization error of a probability P (with compact support) and $n \in \mathbb{N}$ is defined to be

$$\begin{aligned} e_{n,\infty}(P) &= \inf\{\rho_\infty(P,Q)\colon |\text{supp}(Q)| \leq n\} \\ &= \inf\{\text{ess sup}\,\|X - f(X)\|\colon |f(\mathbb{R}^d)| \leq n\} \\ &= \inf_{\substack{\alpha\subset\mathbb{R}^d \\ |\alpha|\leq n}} \sup_{x\in\text{supp}(P)} \min_{a\in\alpha} \|x-a\|, \end{aligned}$$

where ρ_∞ denotes the L_∞–minimal metric, and coincides with the covering radius of the most economical covering of supp(P) by at most n balls of equal radius, that is

$$e_{n,\infty}(P) = \inf_{\substack{\alpha\subset\mathbb{R}^d \\ |\alpha|\leq n}} \min\left\{s \geq 0\colon \bigcup_{a\in\alpha} B(a,s) \supset \text{supp}(P)\right\}.$$

The limit of the sequence $\left(n^{1/d} e_{n,\infty}(P)\right)_{n\geq 1}$ exists in $(0,\infty)$ provided supp(P) is compact Jordan measurable with positive (d-dimensional) volume. The limit

$$Q_\infty(\text{supp}(P)) = \lim_{n\to\infty} n^{1/d} e_{n,\infty}(P)$$

is called covering coefficient or quantization coefficient of order ∞ and can be expressed in terms of the covering coefficient $Q_\infty\left([0,1]^d\right)$ of the unit cube and the volume of supp(P). The results for $r = \infty$ cast new light on the quantization problem for $r < \infty$.

Chapter III deals with the asymptotic behaviour of the quantization error for probabilities P on $\mathbb{R}^d$ which are singular with respect to Lebesgue measure. Following Zador (1982) we introduce the concept of quantization dimension of order r. For $r \in [1, +\infty]$ define

$$e_{n,r}(P) = \begin{cases} V_{n,r}(P)^{1/r} & \text{if } 1 \leq r < \infty, \\ e_{n,\infty}(P) & \text{if } r = \infty. \end{cases}$$

and the quantization dimension of order r

$$D_r(P) = \lim_{n \to \infty} \frac{\log n}{|\log e_{n,r}|}$$

if this limit exists. We compare this concept of dimension to several concepts of dimension which are used in fractal geometry or information theory, like Hausdorff dimension, box dimension, and rate distortion dimension.

Then we consider the class of regular probabilities of dimension D on $\mathbb{R}^d$, where D is a non-negative real number. A probability P on $\mathbb{R}^d$ is regular of dimension D if P has compact support and there is a constant $c > 0$ so that

$$\frac{1}{c} s^D \leq P(B(x,s)) \leq c s^D$$

for all balls $B(x,s)$ whose center lies in the support of P and whose radius s is smaller then a certain value s_0. Examples of this type of measures are the normalized Lebesgue measure on a convex compact set, the normalized surface measure on a convex compact set or a smooth compact manifold, and the normalized Hausdorff measure on certain self-similar sets. For each regular probability P of dimension D the quantization dimension of order $r, r \in [1, +\infty]$, is proved to be D, and moreover

$$0 < \liminf_{n \to \infty} n e_{n,r}(P)^D \leq \limsup_{n \to \infty} n e_{n,r}(P)^D < +\infty.$$

Here $e_{n,r}(P)$ is defined using an arbitrary norm on $\mathbb{R}^d$.

For the l^2-norm on $\mathbb{R}^d$ and P the normalized one-dimensional Hausdorff measure on a rectifiable curve in $\mathbb{R}^d$ of length L we show that the quantization dimension of order $r, r \in [1, +\infty]$, is one and that

$$\lim_{n \to \infty} n e_{n,r}(P) = \begin{cases} Q_r([0,1])^{1/r} L & \text{if } 1 \leq r < +\infty, \\ Q_\infty([0,1]) L & \text{if } r = +\infty, \end{cases}$$

where $Q_r([0,1]) = \frac{1}{2} \frac{1}{1+r}$ and $Q_\infty([0,1]) = \frac{1}{2}$.

Finally we deal with self-similar probabilities on $\mathbb{R}^d$. A probability P on $\mathbb{R}^d$ is self-similar if there are an $N \in \mathbb{N}$, $N \geq 2$, contracting similarity transformations $S_1, \ldots, S_N$ of $\mathbb{R}^d$, and a probability vector $(p_1, \ldots p_N)$ with

$$P = \sum_{i=1}^{N} p_i P \circ S_i^{-1}.$$

P satisfies the strong separation property if the $S_1, \ldots, S_N$ above can be chosen to satisfy $S_i(\text{supp}(P)) \cap S_j(\text{supp}(P)) = \emptyset$ for $i \neq j$. If $(s_1, \ldots, s_N)$ are the contractions numbers corresponding to $(S_1, \ldots, S_N)$ then the similarity dimension is the unique $D \in [0, +\infty)$ with $s_1^D + \ldots + s_N^D = 1$.
If the probability vector $(p_1, \ldots, p_N)$ equals $(s_1^D, \ldots, s_N^D)$ then the corresponding self-similar probability P equals the normalized D-dimensional Hausdorff measure on the support of P. If, in additon, P satisfies the strong separation condition, then the quantization dimension $D_r(P)$ of order r equals D for all $r \in [1, +\infty]$ and, moreover,

$$0 < \liminf_{n\to\infty} n e_{n,r}^D \leq \limsup_{n\to\infty} n e_{n,r}^D < +\infty.$$

If $(p_1, \ldots, p_N) \neq (s_1^D, \ldots, s_N^D)$ and the strong separation condition holds then the quantization dimension $D_r(P)$ of the corresponding P satisfies

$$\sum_{i=1}^{N} (p_i s_i^r)^{\frac{D_r(P)}{r+D_r(P)}} = 1$$

and $D_r(P) < D_t(P)$ if $r < t$. Still, for every $r \in [1, +\infty]$,

$$0 < \liminf_{n\to\infty} n e_{n,r}(P)^{D_r(P)} \leq \limsup_{n\to\infty} n e_{n,r}(P)^{D_r(P)} < +\infty.$$

Thus, self-similar probabilities constitute a class of probabilities for which the quantization dimensions of different orders do not all agree. It remains an open problem for which probabilities $\lim_{n\to\infty} n e_{n,r}(P)^{D_r(P)}$ exists, but it can be shown that for the classical Cantor distribution this limit does not exist if $r = 2$.

In the present book we do not intend to give a complete overview over the large subject of quantization. We will focus on the quantization problem as stated earlier, the so-called fixed rate quantization problem, and develop the underlying theory in a mathematically rigorous way. For a comprehensive recent survey of the theory of quantization including its historical development we refer the reader to the article of Gray and Neuhoff (1998). This article also contains an extensive list of papers published in electrical engineering journals on the subject.

Acknowledgement. Helpful comments by W. Quebbemann and F. Fehringer at an early stage of the project are gratefully acknowledged. Thanks are due to H. Strasser and M. Scheutzow for having invited the authors to give some lectures on quantization at the Wirtschaftsuniversität Wien in May 1997 and the Technische Universität Berlin in August 1997. We further wish to thank S. Stark for his help in preparing the figures and doing the numerical computations.

Chapter I

General properties of the quantization for probability distributions

In this chapter we introduce the quantization problem for probability distributions on $\mathbb{R}^d$ with norm-based distortion measure and derive the basic features of optimal quantizers. The investigation of optimal quantizers requires the concepts of Voronoi diagrams and Voronoi partitions and the concepts of centers and moments of probability distributions on $\mathbb{R}^d$. This chapter also serves to develop the properties of these notions as needed for the quantization problem.

1 Voronoi partitions

Voronoi partitions of $\mathbb{R}^d$ will play a central role as optimal quantizing partitions for probability distributions on $\mathbb{R}^d$. In this section we introduce Voronoi regions, Voronoi diagrams and Voronoi partitions with respect to discrete point sets and describe some of their basic properties.

1.1 General norms

Consider a nonempty subset α of $\mathbb{R}^d$. Throughout α is assumed to be locally finite in the sense that the number of points of α within any bounded subset of $\mathbb{R}^d$ is finite. This implies that α is countable and closed. The quantization problem is associated with finite point sets. However, in Chapter II we deal with lattices which are infinite sets of regulary placed points.

Let $\| \ \|$ denote any norm on $\mathbb{R}^d$. The **Voronoi region** generated by $a \in \alpha$ is defined by

$$W(a|\alpha) = \{x \in \mathbb{R}^d : \|x - a\| = \min_{b \in \alpha} \|x - b\|\} \tag{1.1}$$

and $\{W(a|\alpha): a \in \alpha\}$ is called the **Voronoi diagram** of α; see Figure 1.1. Thus $W(a|\alpha)$ consists of all points x such that a is a nearest point to x in α. The dependence of the Voronoi regions (and of several other objects occuring later) on the norm is not explicitely indicated. Most common norms are l_p-norms given by $\|x\| = (\sum_1^d |x_i|^p)^{1/p}$ for $1 \leq p < \infty$ and $\|x\| = \max_{1 \leq i \leq d} |x_i|$ for $p = \infty$.

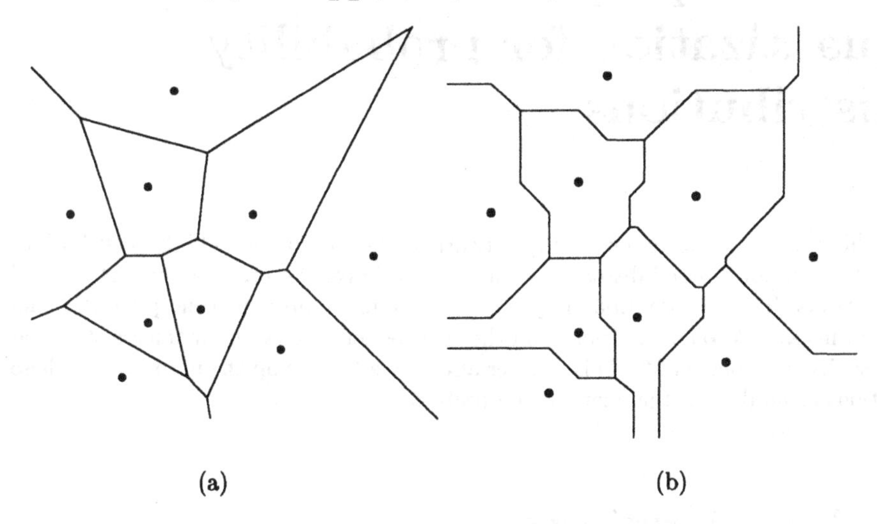

Figure 1.1: Voronoi diagram of a finite set in $\mathbb{R}^2$ with respect to the l_p-norm for (a) $p = 2$ and (b) $p = 1$

A family $\mathcal{A}$ of subsets of $\mathbb{R}^d$ is called **locally finite** if the number of sets in $\mathcal{A}$ intersecting any bounded subset of $\mathbb{R}^d$ is finite. If $x \in \mathbb{R}^d$ and A is a nonempty subset of $\mathbb{R}^d$, the distance from x to A is

$$d(x, A) = \inf_{a \in A} \|x - a\|.$$

The closed ball with center $a \in \mathbb{R}^d$ and radius $r \geq 0$ is denoted by

$$B(a, r) = \{x \in \mathbb{R}^d : \|x - a\| \leq r\}.$$

1.1 Proposition
The Voronoi diagram $\{W(a|\alpha) : a \in \alpha\}$ is a locally finite covering of $\mathbb{R}^d$.

Proof
Let $x \in \mathbb{R}^d$. Since locally finite subsets of $\mathbb{R}^d$ are closed, there exists $a \in \alpha$ such that $\|x - a\| = d(x, \alpha)$ and thus $x \in W(a|\alpha)$. This proves that the Voronoi diagram is a

covering of $\mathbb{R}^d$, that is

$$\bigcup_{a\in\alpha} W(a|\alpha) = \mathbb{R}^d.$$

Moreover, let $\gamma = \{a \in \alpha\colon W(a|\alpha) \cap B(0,s) \neq \emptyset\}$ with $s \geq \min_{a\in\alpha} \|a\|$. Choose $b \in \alpha \cap B(0,s)$. If $a \in \gamma$, then there exists $x \in B(0,s)$ such that $\|x-a\| \leq \|x-b\|$ implying that

$$\|a\| \leq \|x-b\| + \|x\| \leq 2\|x\| + \|b\| \leq 3s.$$

This gives $\gamma \subset B(0,3s)$ and hence, γ is finite. Thus the Voronoi diagram is locally finite. □

The **open Voronoi region** generated by $a \in \alpha$ is defined by

(1.2) $$W_0(a|\alpha) = \{x \in \mathbb{R}^d\colon \|x-a\| < \min_{b\in\alpha\setminus\{a\}} \|x-b\|\}.$$

These regions are pairwise disjoint but do not provide a covering of $\mathbb{R}^d$. A Borel measurable partition $\{A_a : a \in \alpha\}$ of $\mathbb{R}^d$ is called **Voronoi partition** of $\mathbb{R}^d$ with respect to α (and P) if

(1.3) $$A_a \subset W(a|\alpha) \quad (P\text{-a.s.}) \text{ for every } a \in \alpha,$$

where P denotes a Borel probability measure on $\mathbb{R}^d$. Proposition 1.1 shows that Voronoi partitions of $\mathbb{R}^d$ with respect to α do exist and are locally finite. Furthermore, the elements A_a of a Voronoi partition satisfy

$$W_0(a|\alpha) \subset A_a \text{ for every } a \in \alpha.$$

The Voronoi regions are closed and star-shaped relative to their generator point, that is, the line segment joining any $x \in W(a|\alpha)$ and the point a is contained in $W(a|\alpha)$. In case $d = 1$, where the underlying norm is throughout the absolute value, Voronoi regions are closed intervals. For $a, b \in \mathbb{R}^d$, let

(1.4) $$H(a,b) = \{x \in \mathbb{R}^d : \|x-a\| \leq \|x-b\|\}.$$

be the Leibnitz "halfspace". Then $H(a,b) = W(a|\{a,b\})$ and

(1.5) $$W(a|\alpha) = \bigcap_{b\in\alpha} H(a,b).$$

1.2 Proposition

(a) $W(a|\alpha)$ *is closed and star-shaped relative to* a.

(b) $\operatorname{int} W(a|\alpha) = \bigcap_{b\in\alpha} \operatorname{int} H(a,b)$ *and* $W_0(a|\alpha)$ *is an open subset of* $\operatorname{int} W(a|\alpha)$ *which is star-shaped relative to* a. *In particular,* $a \in \operatorname{int} W(a|\alpha)$.

(c) $\partial W(a|\alpha) = \bigcup_{b\in\alpha} \partial H(a,b) \cap W(a|\alpha)$.

Proof

(a) Let $x \in W(a|\alpha)$ and $0 \leq s \leq 1$. The point $y = sx + (1-s)a$ on the line segment joining a and x satisfies

$$\|x - y\| + \|y - a\| = \|x - a\|.$$

Since

$$\|x - a\| \leq \|x - b\| \leq \|x - y\| + \|y - b\| \text{ for every } b \in \alpha,$$

we obtain $y \in W(a|\alpha)$. This shows that $W(a|\alpha)$ is star-shaped relative to a. The Voronoi region is obviously closed.

(b) The region $W_0(a|\alpha) = \{x \in \mathbb{R}^d \colon \|x - a\| < d(x, \alpha \setminus \{a\})\}$ is open because the distance function $d(\cdot, A)$ is continuous. As in the proof of (a) one shows that $W_0(a|\alpha)$ is star-shaped relative to the point a. Furthermore, we have

$$\operatorname{int} W(a|\alpha) \subset \bigcap_{b\in\alpha} \operatorname{int} H(a,b) \subset W(a|\alpha).$$

It remains to show that $\bigcap_{b\in\alpha} \operatorname{int} H(a,b)$ is open. This is clearly true if α is finite. So assume α is not finite. Let $x \in \bigcap_{b\in\alpha} \operatorname{int} H(a,b)$ and set $\gamma = \{b \in \alpha : \|x-a\| = \|x-b\|\}$. Since $\gamma \subset \alpha \cap B(x, \|x - a\|)$, γ is finite by the local finiteness of α. Therefore,

$$\bigcap_{b\in\gamma} \operatorname{int} H(a,b) \ \cap \ W_0(a|(\alpha \setminus \gamma) \cup \{a\})$$

is an open subset of $\bigcap_{b\in\alpha} \operatorname{int} H(a,b)$ containing x.

(c) follows immediately from (b). In fact, we have

$$\begin{aligned}
\partial W(a|\alpha) &= W(a|\alpha) \cap (\operatorname{int} W(a|\alpha))^c \\
&= \bigcup_{b\in\alpha} (\operatorname{int} H(a,b))^c \cap W(a|\alpha) \\
&= \bigcup_{b\in\alpha} \partial H(a,b) \cap W(a|\alpha).
\end{aligned}$$

□

By the preceding proposition, one can find Voronoi partitions of $\mathbb{R}^d$ with respect to α consisting of Borel sets A_a which are star-shaped relative to $a \in \alpha$. In fact, let $\alpha = \{a_1, a_2, \dots\}$ be an enumeration of α and set, for instance,

$$\begin{aligned}
A_1 &= W(a_1|\alpha), \\
A_k &= W(a_k|\alpha) \setminus \bigcup_{j<k} W(a_j|\alpha) \\
&= W(a_k|\alpha) \cap W_0(a_k|\{a_1, \dots, a_k\}), \quad k \geq 2.
\end{aligned}$$

Here ties are broken in favour of smaller indices. Some difficulties arise from the fact that the intersection of different Voronoi regions may have interior points. This corresponds to the fact that the separator of two points may have interior points; see the subsequent Example 1.4. However, if the underlying norm is strictly convex this cannot happen.

For $a, b \in \mathbb{R}^d, a \neq b$, the **separator** is defined by

$$S(a,b) = \{x \in \mathbb{R}^d : \|x-a\| = \|x-b\|\}. \tag{1.6}$$

The separator contains the midpoint $(a+b)/2$ but no other point from the line through a and b. The norm $\|\ \|$ is said to be strictly convex if $\|x\| = \|y\| = 1, x \neq y$ implies $\|sx + (1-s)y\| < 1$ for every $s \in (0,1)$. The l_p-norms are strictly convex for $1 < p < \infty$, while the l_1-norm and the l_∞-norm are not strictly convex.

1.3 Proposition (Strictly convex norms)
Suppose the underlying norm is strictly convex.

(a) $\operatorname{int} W(a|\alpha) = W_0(a|\alpha)$.

(b) $\partial W(a|\alpha) = \bigcup_{\substack{b \in \alpha \\ b \neq a}} S(a,b) \cap W(a|\alpha)$,

(c) $W(a|\alpha) = \operatorname{cl} W_0(a|\alpha)$.

Proof

The proof is based on the following observation. For $b \in \alpha \setminus \{a\}$, we have

$$sx + (1-s)a \in \{y \in \mathbb{R}^d : \|y-a\| < \|y-b\|\} \tag{1.7}$$

for every $x \in H(a,b), 0 \leq s < 1$.

To verify this, let $y = sx + (1-s)a$ with $0 < s < 1$. Notice that $y \in H(a,b)$ by Proposition 1.2 (a) and $\frac{1}{s}(y-a) = x - a$. Assume $\|y-a\| = \|y-b\|$. From the strict convexity of the norm it follows that

$$\|y - a - s(b-a)\| = \|(1-s)(y-a) + s(y-b)\| < \|y-a\|.$$

This gives

$$\begin{aligned}
\|x-b\| &= \|\frac{1}{s}(y-a) - (b-a)\| \\
&= \frac{1}{s}\|y - a - s(b-a)\| \\
&< \frac{1}{s}\|y-a\| = \|x-a\| \leq \|x-b\|,
\end{aligned}$$

a contradiction.

(a) In view of Proposition 1.2 (b) we have to show for $b \in \alpha \setminus \{a\}$

$$\operatorname{int} H(a,b) \subset \{x \in \mathbb{R}^d : \|x-a\| < \|x-b\|\}.$$

Let $x \in \operatorname{int} H(a,b)$ with $x \neq a$ and choose $\varepsilon > 0$ such that $B(x,\varepsilon) \subset H(a,b)$. Let $t = 1 + \varepsilon/\|x-a\|$ and $z = a + t(x-a)$. Then

$$\|z-x\| = \|(t-1)x - (t-1)a\| = \varepsilon$$

implying $z \in H(a,b)$. Since $x = \frac{1}{t}z + (1-\frac{1}{t})a$ and $\frac{1}{t} < 1$, it follows from (1.7) that $\|x-a\| < \|x-b\|$.

(b) follows immediately from (a) and Proposition 1.2 (c).

(c) follows from (1.7). □

The preceding proposition implies

$$\begin{aligned} \operatorname{int} W(a|\alpha) \cap \operatorname{int} W(b|\alpha) &= \emptyset, \ a,b \in \alpha, \ a \neq b, \\ W(a|\alpha) \cap W(b|\alpha) &= \partial W(a|\alpha) \cap \partial W(b|\alpha), \ a,b \in \alpha, \ a \neq b, \\ \partial W(a|\alpha) &= \bigcup_{\substack{b \in \alpha \\ b \neq a}} W(b|\alpha) \cap W(a|\alpha), \ a \in \alpha. \end{aligned} \tag{1.8}$$

provided the underlying norm is strictly convex. The following example shows that all assertions of Proposition 1.3 (and of (1.8)) can fail if $\| \ \|$ is an arbitrary norm.

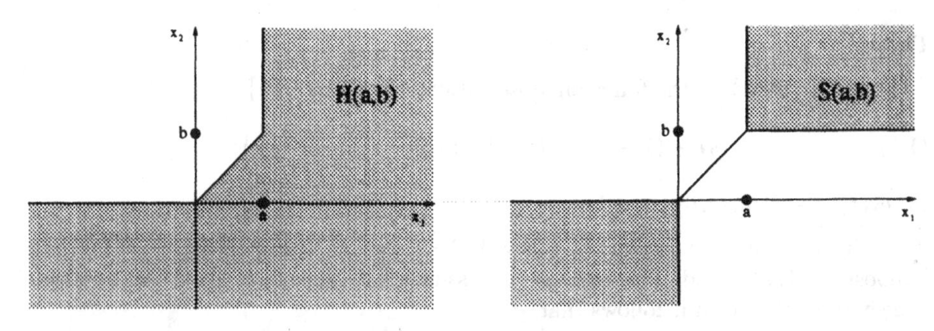

Figure 1.2: Voronoi region and separator with respect to the l_1-norm

1.4 Example

Let the underlying norm on $\mathbb{R}^2$ be the l_1-norm. For $a = (1,0)$ and $b = (0,1)$, we obtain

$$H(a,b) = \{x \in \mathbb{R}^2 : x_2 \leq 0\} \cup \{x \in \mathbb{R}^2 : x_1 \geq 1\} \cup \{x \in \mathbb{R}^2 : x_2 \leq x_1\}$$

and

$$S(a,b) = (-\infty, 0]^2 \cup \{s(1,1) : 0 < s < 1\} \cup [1,\infty)^2.$$

Thus $H(a,b)$ is the union of three halfspaces and the separator is the disjoint union of two quarterspaces and a line segment; see Figure 1.2. Clearly, all assertions of Proposition 1.3 fail for $\alpha = \{a,b\}$.

Under various conditions, Voronoi regions are geometrically regular. Let $|A|$ denote the cardinality of a set A and let λ^d denote the d-dimensional Lebesgue measure.

1.5 Theorem (Boundary theorem)
Each of the following conditions implies

$$\lambda^d(\partial W(a|\alpha)) = 0, \ a \in \alpha.$$

(i) The underlying norm is strictly convex.

(ii) The underlying norm is the l_p-norm with $1 \leq p \leq \infty$.

(iii) $d = 2$.

Proof

According to Proposition 1.2 (c) it is sufficient to show that $\lambda^d(\partial H(a,b)) = 0$ for $a \neq b$. Since $H(a,b) = H(a-b,0)+b$, we may assume without loss of generality that $b = 0$. Set

$$A = H(a,0)^c = \{x \in \mathbb{R}^d : \|x\| < \|x-a\|\}.$$

Then $\partial A = \partial H(a,0)$ and by Proposition 1.2(b), A is open and star-shaped relative to $0 \in A$. For $x \in \mathbb{R}^d$, let

$$I(x) = \{t > 0 : tx \in \partial A\}.$$

Since A and $\mathrm{cl}(A)$ are star-shaped, $I(x)$ is a closed interval (possibly empty, possibly degenerate). Moreover, since

$$\begin{aligned}\lambda^d(\partial A) &= \int_S \int_{\mathbb{R}_+} r^{d-1} 1_{\partial A}(rx)\, dr\, d\sigma(x) \\ &= \int_S \int_{I(x)} r^{d-1}\, dr\, d\sigma(x),\end{aligned}$$

where $S = \{x \in \mathbb{R}^d : \|x\|_{l_2} = 1\}$ denotes the unit l_2-sphere in $\mathbb{R}^d$ and σ its surface measure, we find that $\lambda^d(\partial A) = 0$ if and only if

$$|I(x)| \leq 1 \tag{1.9}$$

for σ-almost all $x \in S$.

Now assume (i). By (1.7), $sy \in A$ for every $y \in \partial A, 0 \leq s < 1$. Therefore condition (1.9) is satisfied for every $x \in \mathbb{R}^d$.

Assume (ii). Let $p = 1$ or $p = \infty$. Then $H(a,0)$ is a finite union of polyhedral sets implying $\lambda^d(\partial H(a,0)) = 0$. If $1 < p < \infty$, then the l_p-norm is strictly convex.

Assume (iii). We show that (1.9) holds for every x. Assume on the contrary that there exists a point x and $t > 1$ such that $\{x, tx\} \subset \partial A$. Since $\partial A \subset S(a,0)$, the convex function $g : \mathbb{R} \to \mathbb{R}$ given by $g(s) = \|x - sa\|$ satisfies

$$g(0) = g(1/t) = g(1) = \|x\|.$$

Hence we get

$$g(s) = \|x\| \text{ for every } s \in [0,1].$$

This yields $\{sx : s \geq 1\} \subset S(a,0)$ and thus

$$\{sx : s \geq r\} \subset S(ra,0) \text{ for every } r \geq 0. \tag{1.10}$$

By assumption there exists a sequence $(y_n)_{n\geq 1}$ in A such that $\lim_{n\to\infty} y_n = tx$. Note that x and a are linearly independent because the separator $S(a,0)$ contains $a/2$ but no other point from the line $\{sa : s \in \mathbb{R}\}$. So, using $d = 2$, we have

$$y_n = s_n a + t_n x\ ,\ s_n, t_n \in \mathbb{R}.$$

Then $s_n \to 0$ and $t_n \to t > 1$. Choose $n_0 \in \mathbb{N}$ such that

$$\begin{aligned} &|s_n| < 1, && t_n \geq -s_n, \\ &0 < \frac{1}{s_n + t_n} < 1, && \frac{t_n}{1 - s_n} \geq 1 \quad \text{for every } n \geq n_0. \end{aligned}$$

We claim that $s_n < 0$ for $n \geq n_0$. We have

$$a + \frac{1}{1-s_n}(y_n - a) = (1 - \frac{1}{1-s_n})a + \frac{s_n}{1-s_n}a + \frac{t_n}{1-s_n}x = \frac{t_n}{1-s_n}x.$$

Therefore, it follows from (1.10) that

$$a + \frac{1}{1-s_n}(y_n - a) \in S(a,0)$$

and hence

$$\frac{1}{1-s_n}(y_n - a) \in S(-a,0).$$

Thus the convex function $h_n : \mathbb{R} \to \mathbb{R}$ given by $h_n(s) = \|y_n - a + sa\|$ satisfies

$$h_n(0) = h_n(1 - s_n) = \|y_n - a\|.$$

This implies

$$\begin{aligned} h_n(s) &\leq \|y_n - a\|\ ,\ 0 \leq s \leq 1 - s_n, \\ h_n(s) &\geq \|y_n - a\|\ ,\ s \geq 1 - s_n. \end{aligned}$$

Since $h_n(1) = \|y_n\| < \|y_n - a\|$, one gets $1 < 1 - s_n, n \geq n_0$ and our claim is proved. Since $t_n \geq -s_n > 0$, it follows from (1.10) that

$$\frac{t_n}{s_n + t_n} x \in S(-\frac{s_n}{s_n + t_n} a, 0), \; n \geq n_0.$$

Hence

$$\|\frac{1}{s_n + t_n} y_n\| = \|\frac{t_n}{s_n + t_n} x + \frac{s_n}{s_n + t_n} a\| = \frac{t_n}{s_n + t_n}\|x\|, \; n \geq n_0.$$

Moreover, again by (1.10) we have

$$\frac{t_n}{s_n + t_n} x \in S(\frac{t_n}{s_n + t_n} a, 0)$$

and therefore

$$\|\frac{1}{s_n + t_n} y_n - a\| = \|\frac{t_n}{s_n + t_n} x - \frac{t_n}{s_n + t_n} a\| = \frac{t_n}{s_n + t_n}\|x\|, \; n \geq n_0.$$

This yields

$$\frac{1}{s_n + t_n} y_n \in S(a, 0), \; n \geq n_0,$$

a contradiction in view of $y_n \in A$ and $0 < 1/(s_n + t_n) < 1$. □

For a Borel subset C of $\mathbb{R}^d$ and a Borel measure μ on $\mathbb{R}^d$, a **μ-tesselation** of C is a countable covering $\{C_n : n \in \mathbb{N}\}$ of C by Borel subsets $C_n \subset C$ such that $\mu(C_n \cap C_m) = 0$ for $n \neq m$. A λ^d-tesselation is simply called **tesselation.** In view of Propositions 1.1 and 1.2 the Voronoi diagram of α is a μ-tesselation of $\mathbb{R}^d$ if and only if

$$\mu(\mathbb{R}^d \setminus \bigcup_{a \in \alpha} W_0(a|\alpha)) = 0;$$

it is a tesselation of $\mathbb{R}^d$ if and only if

$$\text{int}\, W(a|\alpha) \cap \text{int}\, W(b|\alpha) = \emptyset \text{ for every } a, b \in \alpha, \; a \neq b,$$
$$\lambda^d(\partial W(a|\alpha)) = 0 \text{ for every } a \in \alpha.$$

We know from the Example 1.4 that the Voronoi diagram of α, in general does not provide a tesselation of the space $\mathbb{R}^d$. (Notice that the Voronoi diagrams in Figure 1.1 provide tesselations.) According to (1.8) and Theorem 1.5, the Voronoi diagram of α with respect to a strictly convex norm is a tesselation of $\mathbb{R}^d$.

Two further properties of Voronoi regions concerning neighbouring regions and equivariance under similarity transformations are of interest. A bijective mapping $T: \mathbb{R}^d \to \mathbb{R}^d$ is called **similarity transformation** if there exists $c \in (0, \infty)$, the **scaling number**, such that $\|Tx - Ty\| = c\|x - y\|$ for every $x, y \in \mathbb{R}^d$. Let $T(\alpha) = \{Ta : a \in \alpha\}$; $T(\alpha)$ is locally finite.

1.6 Lemma
Let $T : \mathbb{R}^d \to \mathbb{R}^d$ be a similarity transformation. Then

$$W(Ta|T(\alpha)) = TW(a|\alpha).$$

Proof

Obvious. □

Voronoi regions are determined by their neighbouring regions in the following sense.

1.7 Lemma
For $a \in \alpha$, let

$$\beta = \{b \in \alpha : W(b|\alpha) \cap W(a|\alpha) \neq \emptyset\}.$$

Then $W(a|\alpha) = W(a|\beta)$.

Proof

Clearly we have $W(a|\alpha) \subset W(a|\beta)$. To prove the converse inclusion, let $x \in \mathbb{R}^d \setminus W(a|\alpha)$ and consider the line segment $\{y_s : s \in [0,1]\}$ joining a and x, where $y_s = sx + (1-s)a$. Since $W(a|\alpha)$ is closed, $a \in \operatorname{int} W(a|\alpha)$, and $W(a|\alpha)$ is star-shaped relative to a, we obtain

$$W(a|\alpha) \cap \{y_s : s \in [0,1]\} = \{y_s : s \in [0, s_0]\}$$

for some $0 < s_0 < 1$. By the local finiteness of the Voronoi diagram,

$$\gamma = \{c \in \alpha : W(c|\alpha) \cap \{y_s : s \in (s_0, 1]\} \neq \emptyset\}$$

is finite. Since $\bigcup_{c \in \gamma} W(c|\alpha)$ is closed, one gets

$$\{y_s : s \in [s_0, 1]\} \subset \bigcup_{c \in \gamma} W(c|\alpha).$$

Choose $b \in \gamma$ with $y_{s_0} \in W(b|\alpha)$. Then $b \in \beta$ and $y_t \in W(b|\alpha)$ for some $t \in (s_0, 1]$. Since $y_t \notin W(a|\alpha)$, the point y_t satisfies $y_t \notin W(a|\beta)$. This implies $x \notin W(a|\beta)$. □

Notice that bounded Voronoi regions have a finite number of neighbouring regions by the local finiteness of Voronoi diagrams.

1.2 Euclidean norms

Voronoi regions with respect to euclidean norms exhibit some special features. Let $\langle\,,\rangle$ be any scalar product on $\mathbb{R}^d$ and $\|x\| = \langle x, x\rangle^{1/2}$. Then for $a \neq b, H(a,b)$ is the closed halfspace

$$H(a,b) = \{x \in \mathbb{R}^d : \langle a - b, x - \frac{1}{2}(a+b)\rangle \geq 0\} \tag{1.11}$$

bounded by the separating hyperplane

$$S(a,b) = \{x \in \mathbb{R}^d : \langle a-b, x-\frac{1}{2}(a+b)\rangle = 0\}. \tag{1.12}$$

The hyperplane $S(a,b)$ contains the midpoint $(a+b)/2$ and is perpendicular (with respect to $\langle\ ,\ \rangle$) to the line through a and b. Thus the Voronoi regions $W(a|\alpha)$ are convex. If α is finite, then $W(a|\alpha)$ is a polyhedral set, that is, a finite intersection of closed halfspaces in $\mathbb{R}^d$. In the sequel a (convex) polytope means a compact polyhedral set. By Lemma 1.7, bounded Voronoi regions are polytopes. The following example shows that, in general, unbounded Voronoi regions are not polyhedral sets.

1.8 Example
Let the norm on $\mathbb{R}^2$ be the l_2-norm. Consider the set $\alpha = \{a_n : n \geq 0\}$ with $a_0 = (2,0)$ and $a_n = (0,n)$ for $n \geq 1$. Then α is locally finite and the points

$$(\frac{1}{4}(n^2+n)+1, n+\frac{1}{2}),\ n \geq 1$$

are extreme points of $W(a_0|\alpha)$; see Figure 1.3. Since polyhedral sets in $\mathbb{R}^d$ have a finite number of extreme points, $W(a_0|\alpha)$ is not polyhedral.

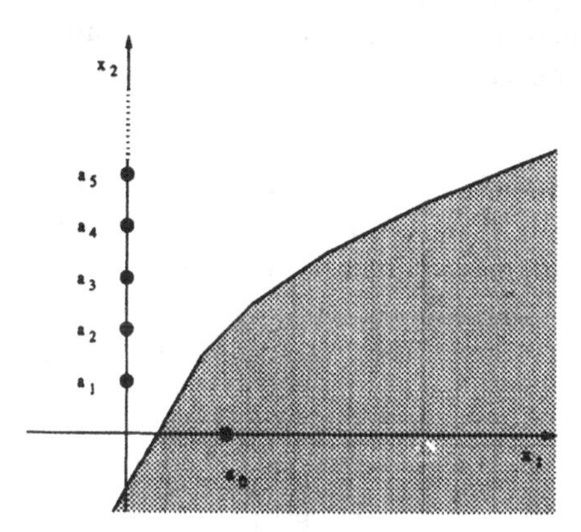

Figure 1.3: Voronoi region with respect to l_2-norm which is not a polyhedral set

1.9 Remark
As indicated above, the Voronoi regions are convex in the euclidean case. It is an interesting fact that the convexity of Voronoi regions even characterizes euclidean norms. More precisely, if $W(a|\alpha)$ is convex for every finite subset α of $\mathbb{R}^d$ and every $a \in \alpha$, then the underlying norm is euclidean. This is a classical result of Mann (1935). See also Gruber (1974).

For euclidean norms, there is a simple characterization of boundedness of Voronoi regions. Denote by $\operatorname{conv}\alpha$ the convex hull of α.

1.10 Proposition
Let $\|x\| = \langle x, x\rangle^{1/2}$ for some scalar product $\langle\, ,\, \rangle$ on $\mathbb{R}^d$ and let $a \in \alpha$. Then $W(a|\alpha)$ is bounded if and only if $a \in \operatorname{int} \operatorname{conv} \alpha$.

Proof
For $u \in \mathbb{R}^d, u \neq 0$, consider the halfline $L_u = \{a + su : s \geq 0\}$ with initial point a. We have $L_u \subset W(a|\alpha)$, that is,

$$\begin{aligned}\|a + su - b\|^2 &= \|a - b\|^2 + 2s\langle u, a - b\rangle + s^2\|u\|^2 \\ &\geq \|a + su - a\|^2 = s^2\|u\|^2 \text{ for every } b \in \alpha,\ s \geq 0\end{aligned}$$

if and only if $\langle u, a\rangle \geq \langle u, b\rangle$ for every $b \in \alpha$.

Assume $a \in \partial \operatorname{conv} \alpha$. Then there passes a support hyperplane to $\operatorname{conv} \alpha$ through a, that is, $\langle u, a\rangle \geq \langle u, b\rangle$ for every $b \in \alpha$ and some $u \in \mathbb{R}^d, u \neq 0$ (cf. Webster, 1994, Theorem 2.4.12). Hence, $L_u \subset W(a|\alpha)$ so that $W(a|\alpha)$ is not bounded. Now assume $a \in \operatorname{int} \operatorname{conv} \alpha$. Then for every $u \in \mathbb{R}^d, u \neq 0$ there exists $b \in \alpha$ such that $\langle u, a\rangle < \langle u, b\rangle$. Therefore, $W(a|\alpha)$ does not contain any halfline with initial point a. This implies the boundedness of $W(a|\alpha)$ (cf. Webster, 1994, Theorem 2.5.1). □

Notice that for arbitrary norms the interior point condition for the generator point does not imply the boundedness of the corresponding Voronoi region. This is illustrated by the following example. See also Figure 1.1(b).

1.11 Example
Let the underlying norm on $\mathbb{R}^2$ be the l_1-norm. Consider $\alpha = \{(0,0), (0,-1), (2,1), (-2,1)\}$ and let $a = (0,0)$. Then $a \in \operatorname{int} \operatorname{conv} \alpha$, but $W(a|\alpha)$ is unbounded since, for instance, the halfline $\{s(0,1) : s \geq 0\}$ is contained in $W(a|\alpha)$; see Figure 1.4.

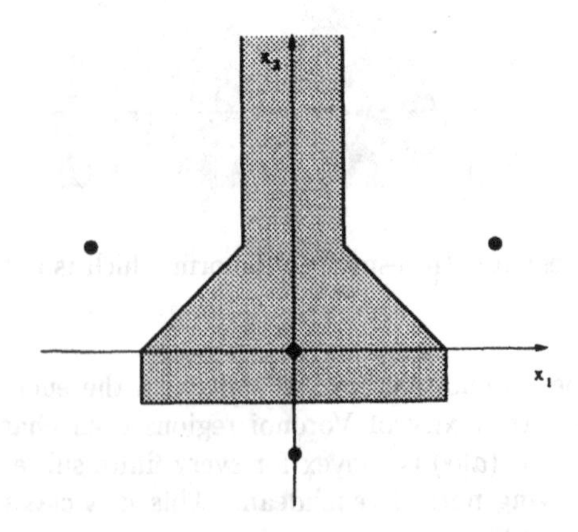

Figure 1.4: Unbounded Voronoi region with respect to the l_1-norm generated by an interior point of $\operatorname{conv} \alpha$

Notes

For detailed treatments of Voronoi diagrams of finite point sets we refer to the notable book by Okabe et al. (1992), the review article by Aurenhammer (1991), and the book by Klein (1989). A discussion of random Voronoi tesselations with respect to the l_2-norm may be found in Møller (1994). Theorem 1.5 (iii) on the geometric regularity of Voronoi regions is certainly known but we are not aware of a reference.

1.12 Conjecture
The assertion of Theorem 1.5, that is, $\lambda^d(\partial W(a|\alpha)) = 0$ for every $a \in \alpha$, holds for arbitrary norms and arbitrary dimensions.

2 Centers and moments of probability distributions

We present some facts about centers and moments of probability distributions on $\mathbb{R}^d$ needed in the sequel.

2.1 Uniqueness and characterization of centers

Let $X = (X_1, \dots, X_d)$ be a $\mathbb{R}^d$-valued random variable with distribution P. As before, $\| \ \|$ denotes any norm on $\mathbb{R}^d$. Let $1 \le r < \infty$ and assume throughout

$$E\|X\|^r < \infty.$$

By a **center of P of order r** we mean a point $a \in \mathbb{R}^d$ such that

$$E\|X - a\|^r = \inf_{b \in \mathbb{R}^d} E\|X - b\|^r. \tag{2.1}$$

Let $C_r(P)$ denote the set of all centers of P of order r. Centers of order 1 are usually called (spatial) medians. The **r-th (absolute) moment of P** about the center is defined by

$$V_r(P) = \inf_{a \in \mathbb{R}^d} E\|X - a\|^r. \tag{2.2}$$

We also write $V_r(X)$ and $C_r(X)$ instead of $V_r(P)$ and $C_r(P)$.

For $A \in \mathcal{B}(\mathbb{R}^d)$ bounded with $\lambda^d(A) > 0$, define the **normalized r-th moment of A** about the center by

$$M_r(A) = \frac{V_r(U(A))}{\lambda^d(A)^{r/d}}, \tag{2.3}$$

where $U(A)$ denotes the uniform distribution on A. Normalization yields an important scaling invariance property.

Recall that bijective isometries $T : \mathbb{R}^d \to \mathbb{R}^d$ with $T(0) = 0$ are linear (cf. Semadeni, 1971, Lemma 7.8.3) and λ^d is invariant under bijective isometries on $\mathbb{R}^d$.

2.1 Lemma
Let $T : \mathbb{R}^d \to \mathbb{R}^d$ be a similarity transformation with scaling number $c > 0$.

(a) $C_r(T(X)) = TC_r(X)$,
$V_r(T(X)) = c^r V_r(X)$.

(b) $M_r(T(A)) = M_r(A)$, *if $A \in \mathcal{B}(\mathbb{R}^d)$ is bounded with $\lambda^d(A) > 0$.*

Proof

(a) is obvious.

(b) If X is $U(A)$-distributed, then $T(X)$ is $U(T(A))$-distributed. From (a) it follows that

$$M_r(T(A)) = \frac{V_r(T(X))}{\lambda^d(T(A))^{r/d}} = \frac{c^r V_r(X)}{(c^d \lambda^d(A))^{r/d}} = M_r(A).$$

□

Centers of order r are global minima of the function

$$\psi_r : \mathbb{R}^d \to \mathbb{R}_+,\ \psi_r(a) = E\|X-a\|^r. \tag{2.4}$$

The function ψ_r is obviously convex and hence continuous. Therefore, local minima of ψ_r are global minima of ψ_r.

2.2 Lemma

The level set $\{\psi_r \leq c\}$ is convex and compact for every $c \in \mathbb{R}_+$. In particular, $C_r(P)$ is convex compact and nonempty. Moreover,

$$C_r(P) \subset B(0, 2(2E\|X\|^r)^{1/r})$$

and

$$C_1(P) \subset B(0, 2E\|X\|).$$

Proof

By convexity and continuity of ψ_r, the level sets are convex and closed. Since

$$\|a\| \leq \|x-a\| + \|x\| \leq 2\max\{\|x-a\|, \|x\|\}$$

and hence

$$\begin{aligned}\|a\|^r &\leq 2^r \max\{\|x-a\|^r, \|x\|^r\}\\ &\leq 2^r(\|x-a\|^r + \|x\|^r),\ a, x \in \mathbb{R}^d\end{aligned}$$

one obtains for $a \in \{\psi_r \leq c\}$

$$\|a\|^r \leq 2^r(c + E\|X\|^r).$$

Therefore,

$$\{\psi_r \leq c\} \subset B(0, 2(c+E\|X\|^r)^{1/r}).$$

In case $r = 1$, we have

$$\{\psi_1 \leq c\} \subset B(0,\ c + E\|X\|)).$$

Choosing $c = E\|X\|^r$ and $c = V_r(P)$, respectively, gives the assertions. □

2.3 Example
(a) If P is symmetric (about the origin), then $0 \in C_r(P)$ and thus, $V_r(P) = E\|X\|^r$. In fact, $C_r(P)$ is symmetric by Lemma 2.1 (a) and from convexity of $C_r(P)$ follows $0 \in C_r(P)$.

(b) For the l_2-norm, we obtain $C_2(X) = \{EX\}$ and $V_2(X) = \sum_{i=1}^d \operatorname{Var} X_i$, where $\operatorname{Var} X_i$ denotes the variance of X_i. If the underlying norm is the l_1-norm, then $C_1(X) = \times_1^d \operatorname{Med}(X_i)$, where $\operatorname{Med}(X_i)$ is the set of medians of the real random variable X_i.

The center of a probability distribution need not be unique; think of the median of one-dimensional distributions. Conditions for the uniqueness of the center are derived in the following theorem. Condition (iii) is due to Milasevic and Ducharme (1987) (for euclidean norms) and Kemperman (1987).

2.4 Theorem (Uniqueness)
Each of the following conditions implies $|C_r(P)| = 1$.

(i) *The underlying norm is strictly convex and* $r > 1$.

(ii) $P(S(a,b)) < 1$ *for every* $a, b \in \mathbb{R}^d, a \neq b$, *and* $r > 1$.

(iii) *The underlying norm is strictly convex,* $r = 1$, *and* $P(L) < 1$ *for every line* $L \subset \mathbb{R}^d$.

Proof

We show that ψ_r is strictly convex. This yields the assertion. Let $a, b \in \mathbb{R}^d, a \neq b$, and $0 < s < 1$. Then

$$\begin{aligned}\|x - (sa + (1-s)b)\|^r &= \|s(x-a) + (1-s)(x-b)\|^r \\ &\leq (s\|x-a\| + (1-s)\|x-b\|)^r \\ &\leq s\|x-a\|^r + (1-s)\|x-b\|^r, \quad x \in \mathbb{R}^d.\end{aligned}$$

Let

$$A = \{x \in \mathbb{R}^d : \|s(x-a) + (1-s)(x-b)\|^r = s\|x-a\|^r + (1-s)\|x-b\|^r\}.$$

In case $r > 1$, A is contained in the separator $S(a,b)$ since $t \mapsto t^r$ is strictly convex. If, additionally, the norm is strictly convex, then $A = \emptyset$.

If $r = 1$ and the norm is strictly convex, then A is a subset of the line L through a and b given by $L = \{ta + (1-t)b : t \in \mathbb{R}\}$. To see this, let $x \in A$ and assume $x \neq b$. By strict convexity of the norm, there exists $t \in \mathbb{R}_+$ such that $s(x-a) = t(1-s)(x-b)$. This gives

$$x = \frac{s}{s-t+st}a + \frac{ts-t}{s-t+st}b$$

and hence $x \in L$. Notice that $s - t + st \neq 0$ since otherwise, $a = b$.

Therefore, under any of the above conditions we have $P(A) < 1$. This implies $\psi_r(sa + (1-s)b) < s\,\psi_r(a) + (1-s)\psi_r(b)$. □

Next, we characterize the centers of P by means of the derivative of the underlying norm. Let $\nabla_+\|\ \|(x, y)$ denote the one-sided directional derivative of the norm at $x \in \mathbb{R}^d$ in direction $y \in \mathbb{R}^d$ given by

$$\nabla_+\|\ \|(x,y) = \lim_{t \to 0+} \frac{\|x + ty\| - \|x\|}{t}.$$

The norm $\|\ \|$ is said to be smooth if it is differentiable at every point $x \neq 0$. In the smooth case, the (two-sided) directional derivative exists at every $x \neq 0$ and coincides with $\langle \nabla\|\ \|(x), y\rangle$, that is,

$$\langle \nabla\|\ \|(x), y\rangle = \lim_{t \to 0} \frac{\|x + ty\| - \|x\|}{t}, \quad y \in \mathbb{R}^d,$$

where $\nabla\|\ \|(x)$ denotes the derivative of the norm at x and $\langle x, y\rangle = \sum_1^d x_i y_i$.

The l_p-norms are smooth for $1 < p < \infty$ while the l_1-norm and l_∞-norm are not smooth.

2.5 Lemma
For $a \in \mathbb{R}^d$, we have $a \in C_r(P)$ if and only if

$$\int \|x - a\|^{r-1} \nabla_+\|\ \|(a - x, y)\, dP(x) \geq 0$$

for every $y \in \mathbb{R}^d$. If the underlying norm is smooth, this condition takes the form

$$\int_{\{x \neq a\}} \|x - a\|^{r-1} \nabla\|\ \|(a - x)\, dP(x) = 0, \ r > 1,$$

$$\left| \left\langle \int_{\{x \neq a\}} \nabla\|\ \|(a - x)\, dP(x), y \right\rangle \right| \leq P(\{a\})\|y\|$$

for every $y \in \mathbb{R}^d, r = 1$. (Recall $\langle x, y\rangle = \sum_1^d x_i y_i$.)

Proof

Notice first that $a \in C_r(P)$ if and only if

$$\nabla_+\psi_r(a, y) \geq 0$$

for every $y \in \mathbb{R}^d$. This is a consequence of the convexity of ψ_r. Furthermore,

$$\frac{\psi_r(a + ty) - \psi_r(a)}{t} = \int \frac{\|a - x + ty\|^r - \|a - x\|^r}{t}\, dP(x).$$

The function $g: \mathbf{R} \to \mathbf{R}$ given by $g(t) = \|a - x + ty\|^r$ is convex and thus satisfies

$$g(0) - g(-1) \leq \frac{g(t) - g(0)}{t} \leq g(1) - g(0),\ 0 < t \leq 1$$

(cf. Webster, 1994, Theorem 5.1.1). Therefore, by Lebesgue's dominated convergence theorem

$$\nabla_+ \psi_r(a, y) = \int \nabla_+ \| \ \|^r (a - x, y)\, dP(x).$$

Since

$$\nabla_+ \| \ \|^r (x, y) = r\|x\|^{r-1} \nabla_+ \| \ \| (x, y)$$

this yields the first assertion.

Now assume that the norm is smooth. Since $\nabla_+ \| \ \|(0, y) = \|y\|$ one gets

$$\begin{aligned} &\int \|x - a\|^{r-1} \nabla_+ \| \ \|(a - x, y)\, dP(x) \\ &\quad = \Big\langle \int_{\{x \neq a\}} \|x - a\|^{r-1} \nabla \| \ \|(a - x)\, dP(x), y \Big\rangle + \int_{\{x = a\}} \|x - a\|^{r-1} \|y\|\, dP(x) \end{aligned}$$

for every $y \in \mathbf{R}^d$. This yields the second assertion. □

Remark

(a) Using the dual norm of $\|x\|$ given by

$$\|x\|_D = \sup\{\langle x, y \rangle : \|y\| \leq 1\},$$

the above equivalent condition for $a \in \mathbf{R}^d$ to belong to $C_1(P)$ in the smooth case means

$$\Big\| \int_{\{x \neq a\}} \nabla \| \ \|(a - x)\, dP(x) \Big\|_D \leq P(\{a\}).$$

(b) Suppose the underlying norm is smooth. Then ψ_r is differentiable on $\mathbf{R}^d$ for $r > 1$ while ψ_1 is differentiable at every point $a \in \mathbf{R}^d$ with $P(\{a\}) = 0$. The derivative is given by

$$\nabla \psi_r(a) = r \int_{\{x \neq a\}} \|x - a\|^{r-1} \nabla \| \ \|(a - x)\, dP(x).$$

The diameter of a nonempty bounded subset A of $\mathbf{R}^d$ is the number

$$\mathrm{diam}(A) = \sup\{\|a - b\| : a, b \in A\}.$$

Denote by $\mathrm{supp}(P)$ the topological support of P.

2.6 Lemma

(a) (Euclidean norms) Let $\|x\| = \langle x, x\rangle^{1/2}$ *for some scalar product* $\langle\ ,\ \rangle$ *on* $\mathbb{R}^d$. *Then*

$$C_r(P) \subset \operatorname{cl}\operatorname{conv}(\operatorname{supp}(P)).$$

(b) Suppose $\operatorname{supp}(P)$ *is compact. Then*

$$\sup_{a \in C_r(P)} \min_{x \in \operatorname{supp}(P)} \|x - a\| \leq \operatorname{diam}(\operatorname{supp}(P)).$$

Proof

(a) Let $K = \operatorname{cl}\operatorname{conv}(\operatorname{supp}(P))$ and $a \notin K$. Then K and a can be strictly separated by a hyperplane H, that is , K and a lie on opposite open halfspaces determined by H. Denote by b the orthogonal projection of a onto H. For $x \in K$, let y be the point on the line segment joining a and x which lies in H. Since $\langle y - b, a - b\rangle = 0$, we obtain

$$\|y - a\|^2 = \|y - b\|^2 + \|b - a\|^2 > \|y - b\|^2$$

and therefore,

$$\begin{aligned} \|x - b\| &\leq \|x - y\| + \|y - b\| \\ &< \|x - y\| + \|y - a\| = \|x - a\|. \end{aligned}$$

This implies $a \notin C_r(P)$.

(b) Let $a \in \mathbb{R}^d$ such that $d(a, \operatorname{supp}(P)) > \operatorname{diam}(\operatorname{supp}(P))$ and let $y \in \operatorname{supp}(P)$. Then

$$\|x - y\| \leq \operatorname{diam}(\operatorname{supp}(P)) < \|x - a\|$$

for every $x \in \operatorname{supp}(P)$. This implies $a \notin C_r(P)$ □

The assertion of part (a) of the preceding lemma can fail if $\|\ \|$ is an arbitrary norm. This is exhibited by the following example.

2.7 Example

Let the underlying norm on $\mathbb{R}^2$ be the l_∞-norm. Consider $P = \frac{1}{2}(\delta_{(-1,0)} + \delta_{(1,0)})$, where δ_x is the point mass at x. Since P is symmetric about $EX = (0,0)$, this point belongs to $C_r(P)$ and thus, $V_r(P) = E\|X\|^r = 1$ for every $1 \leq r < \infty$. We find

$$\begin{aligned} C_1(P) &= \{\psi_1 = 1\} = \{x \in \mathbb{R}^2 : |x_1| + |x_2| \leq 1\}, \\ C_r(P) &= \{\psi_r = 1\} = \{s(0,1) : -1 \leq s \leq 1\} \text{ for } r > 1; \end{aligned}$$

see Figure 2.1. Clearly, the assertion of Lemma 2.6 does not hold for P.

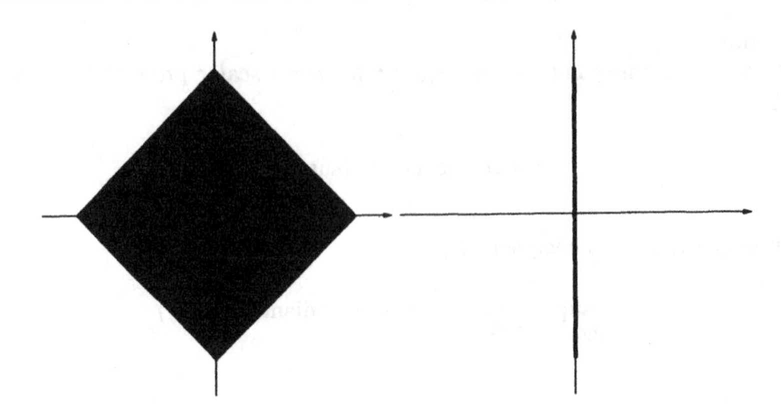

Figure 2.1: $C_1(P)$ and $C_r(P), r > 1$, with respect to the l_∞-norm for a discrete probability P with two supporting points

2.2 Moments of balls

Balls have minimal moments for measures μ which vanish on spheres, i. e. $\mu(\partial B(a,s)) = 0$ for every $a \in \mathbb{R}^d$ and every $s \geq 0$. (Note that $\partial B(a,s) = \{x \in \mathbb{R}^d : \|x - a\| = s\}$.) This statement is meant in the following sense.

2.8 Lemma

Let μ be a Borel measure on $\mathbb{R}^d$ that is finite on compact sets and vanishes on spheres. Then, for every bounded set $A \in \mathcal{B}(\mathbb{R}^d)$ with $\mu(A) > 0$ and every $a \in \mathbb{R}^d$ there is an $s \geq 0$ with $\mu(B(a,s)) = \mu(A)$. Moreover, for such an s,

$$\int_A \|x - a\|^r \, d\mu(x) \geq \int_{B(a,s)} \|x - a\|^r \, d\mu(x).$$

In particular, we have for $a \in C_r(\mu(\cdot|A))$

$$V_r(\mu(\cdot|A)) \geq V_r(\mu(\cdot|B(a,s)),$$

where $\mu(\cdot|A) = \mu(\cdot \cap A)/\mu(A)$.

Proof

Since A is bounded there exists an $s_0 > 0$ with $A \subset B(a, s_0)$, hence $0 < \mu(A) \leq \mu(B(a, s_0)) < \infty$. Since the map $\mathbb{R}_+ \to \mathbb{R}_+, s \mapsto \mu(B(a,s))$ is continuous under the assumptions for μ the intermediate value theorem yields the existence of an $s > 0$ with $s \leq s_0$ and $\mu(B(a,s)) = \mu(A)$. Then $\mu(A \setminus B(a,s)) = \mu(B(a,s) \setminus A)$ and we have

$$\int_A \|x-a\|^r \, d\mu(x) = \int_{B(a,s)} \|x-a\|^r \, d\mu(x) + \int_{A\setminus B(a,s)} \|x-a\|^r \, d\mu(x)$$
$$- \int_{B(a,s)\setminus A} \|x-a\|^r \, d\mu(x).$$

Obviously

$$\int_{A\setminus B(a,s)} \|x-a\|^r \, d\mu(x) \geq s^r \mu(A \setminus B(a,s))$$

and

$$\int_{B(a,s)\setminus A} \|x-a\|^r \, d\mu(x) \leq s^r \mu(B(a,s) \setminus A).$$

This implies

$$\int_{A\setminus B(a,s)} \|x-a\|^r \, d\mu(x) - \int_{B(a,s)\setminus A} \|x-a\|^r \, d\mu(x)$$
$$\geq s^r(\mu(A \setminus B(a,s)) - \mu(B(a,s) \setminus A)) = 0.$$

Hence, the lemma is proved. □

We can deduce a well known fact about the moments of unifom distributions on balls.

2.9 Lemma

We have

$$M_r(B(0,1)) = \min\{M_r(A) \colon A \in \mathcal{B}(\mathbb{R}^d) \text{ bounded}, \lambda^d(A) > 0\}$$

and $B(0,1)$ is the essentially unique minimizer of M_r in that any bounded set $A \in \mathcal{B}(\mathbb{R}^d)$ with $M_r(A) = M_r(B(0,1))$, $\lambda^d(A) = \lambda^d(B(0,1))$, and $0 \in C_r(U(A))$ satisfies

$$\lambda^d(A \,\Delta\, B(0,1)) = 0.$$

If, additionally, A is regularly closed (that is, $A = \mathrm{cl}(\mathrm{int}\, A)$), then $A = B(0,1)$. Moreover,

$$M_r(B(0,1)) = \frac{d}{(d+r)\lambda^d(B(0,1))^{r/d}}. \tag{2.5}$$

Proof

The first assertion follows from Lemma 2.8 with the choice $\mu = \lambda^d$ and Lemma 2.1 (b). As for uniqueness, let A be a set with the above properties. Then

$$\int_A \|x\|^r \, dx = \int_{B(0,1)} \|x\|^r \, dx.$$

It follows that

$$\lambda^d(B(0,1) \setminus A) = \lambda^d(A \setminus B(0,1)) \leq \int_{A \setminus B(0,1)} \|x\|^r \, dx$$
$$= \int_{B(0,1) \setminus A} \|x\|^r \, dx \leq \lambda^d(B(0,1) \setminus A).$$

Therefore,

$$\int_{A \setminus B(0,1)} (\|x\|^r - 1) \, dx = 0$$

which implies $A \setminus B(0,1) \subset \partial B(0,1)$ λ^d-a.s. Since $\lambda^d(\partial B(0,1)) = 0$, we obtain $\lambda^d(A \,\Delta\, B(0,1)) = 0$. If A is closed, then $B(0,1) \subset A$. Otherwise $\{\|x\| < 1\} \setminus A$ is a nonempty open set implying

$$\lambda^d(B(0,1) \setminus A) \geq \lambda^d(\{\|x\| < 1\} \setminus A) > 0,$$

a contradiction. It follows that $\{\|x\| < 1\} \subset \operatorname{int} A$ and hence

$$\lambda^d(B(0,1)) = \lambda^d(\{\|x\| < 1\}) \leq \lambda^d(\operatorname{int} A) \leq \lambda^d(A).$$

Therefore, $\lambda^d(\{\|x\| < 1\}) = \lambda^d(\operatorname{int} A)$ which implies $\{\|x\| < 1\} = \operatorname{int} A$. From regular closedness of A follows $A = B(0,1)$.

Moreover, in view of the symmetry of $U(B(0,1))$ one gets

$$V_r(U(B(0,1))) = \int \|x\|^r \, dU(B(0,1))(x)$$
$$= \int_0^\infty U(B(0,1))(\|x\|^r > t) \, dt$$
$$= \int_0^1 (1 - t^{d/r}) \, dt = \frac{d}{d+r}.$$

This gives the formula (2.5). □

In view of the above formula for $M_r(B(0,1))$ it is worth to recall that the volume of unit balls with respect to the l_p-norms for $1 \leq p < \infty$ is given by

$$\lambda^d(B(0,1)) = \frac{(2\Gamma(1 + \frac{1}{p}))^d}{\Gamma(1 + \frac{d}{p})} \tag{2.6}$$

(cf. e. g. Pisier, 1989, p. 11).

Notes

Among spatial centers the spatial medians have received special attention. We refer to the survey article by Small (1990) for a discussion of several notions of spatial medians. A good source for norm-based medians as defined in (2.1) is Kemperman (1987).

3 The quantization problem

In this section we will give several equivalent formulations of the quantization problem for probability distributions on $\mathbb{R}^d$ with norm-based distortion measure. Let X denote a $\mathbb{R}^d$-valued random variable with distribution P. For $n \in \mathbb{N}$, let $\mathcal{F}_n$ be the set of all Borel measurable maps $f : \mathbb{R}^d \to \mathbb{R}^d$ with $|f(\mathbb{R}^d)| \leq n$. The elements of $\mathcal{F}_n$ are called **n-quantizers**. For each $f \in \mathcal{F}_n$, $f(X)$ gives a quantized version of X. Let $1 \leq r < \infty$ and assume

$$E\|X\|^r < \infty.$$

The **n-th quantization error for P of order r** is defined by

$$V_{n,r}(P) = \inf_{f \in \mathcal{F}_n} E\|X - f(X)\|^r. \tag{3.1}$$

We will also write $V_{n,r}(X)$ instead of $V_{n,r}(P)$. A quantizer $f \in \mathcal{F}_n$ is called **n-optimal for P of order r** if

$$V_{n,r}(P) = E\|X - f(X)\|^r.$$

Note that $V_{1,r}(P) = V_r(P)$.

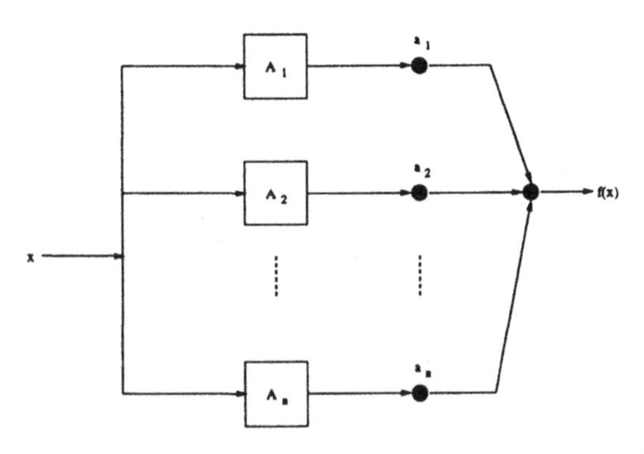

Figure 3.1: Quantization scheme

For fixed $n \in \mathbb{N}$, searching for an n-optimal quantizer is equivalent to the n-centers problem.

3.1 Lemma

$$V_{n,r}(P) = \inf_{\substack{\alpha \subset \mathbb{R}^d \\ |\alpha| \leq n}} E \min_{a \in \alpha} \|X - a\|^r.$$

Proof

For $f \in \mathcal{F}_n$, let $\alpha = f(\mathbb{R}^d)$ and $A_a = \{f = a\}$, $a \in \alpha$. Then

$$\begin{aligned} E\|X - f(X)\|^r &= \sum_{a\in\alpha} \int_{A_a} \|x - a\|^r \, dP(x) \\ &\geq \sum_{a\in\alpha} \int_{A_a} \min_{b\in\alpha} \|x - b\|^r \, dP(x) \\ &= E \min_{b\in\alpha} \|X - b\|^r . \end{aligned}$$

Conversely, for $\alpha \subset \mathbb{R}^d$ with $|\alpha| \leq n$, let $\{A_a : a \in \alpha\}$ be a Voronoi partition of $\mathbb{R}^d$ with respect to α and let $f = \sum_{a\in\alpha} a 1_{A_a}$. Then $f \in \mathcal{F}_n$ and

$$E \min_{a\in\alpha} \|X - a\|^r = \sum_{a\in\alpha} \int_{A_a} \|x - a\|^r \, dP(x) = E\|X - f(X)\|^r .$$

□

A set $\alpha \subset \mathbb{R}^d$ with $|\alpha| \leq n$ is called **n-optimal set of centers for P of order** r if

$$V_{n,r}(P) = E \min_{a\in\alpha} \|X - a\|^r .$$

The proof of Lemma 3.1 shows that if f is an n-optimal quantizer, then $f(\mathbb{R}^d)$ is an n-optimal set of centers. Conversely, if $\alpha \subset \mathbb{R}^d$ is an n-optimal set of centers and $\{A_a : a \in \alpha\}$ is a Voronoi partition of $\mathbb{R}^d$ with respect to α, then $f = \sum_{a\in\alpha} a 1_{A_a}$ is an n-optimal quantizer. Recall that by Proposition 1.2, the sets A_a may be chosen to be star-shaped relative to a.

Let $C_{n,r}(P)$ denote the set of all n-optimal sets of centers for P of order r. We also write $C_{n,r}(X)$ instead of $C_{n,r}(P)$. Note that $C_{1,r}(P)$ can be identified with $C_r(P)$.

For $A \in \mathcal{B}(\mathbb{R}^d)$ bounded with $\lambda^d(A) > 0$, define the **normalized n-th quantization error for A of order r** by

$$M_{n,r}(A) = \frac{V_{n,r}(U(A))}{\lambda^d(A)^{r/d}}. \tag{3.2}$$

The following equivariance, scaling and invariance properties extend those of Lemma 2.1.

3.2 Lemma
Let $T\colon \mathbb{R}^d \to \mathbb{R}^d$ be a similarity transformation with scaling number $c > 0$.

(a) $C_{n,r}(T(X)) = TC_{n,r}(X)$,
$V_{n,r}(T(X)) = c^r V_{n,r}(X)$.

(b) $M_{n,r}(T(A)) = M_{n,r}(A)$, *if* $A \in \mathcal{B}(\mathbb{R}^d)$ *is bounded with* $\lambda^d(A) > 0$.

Proof

Obvious. □

Next, we show that the quantization problem is equivalent to a partitioning problem for the space $\mathbb{R}^d$.

3.3 Lemma

$$V_{n,r}(P) = \inf_{\mathcal{A}} \sum_{A \in \mathcal{A}} V_r(P(\cdot|A))P(A),$$

where the infimum is taken over all Borel measurable partitions $\mathcal{A}$ *of* $\mathbb{R}^d$ *with* $|\mathcal{A}| \leq n$.

Proof

For $f \in \mathcal{F}$, let $\alpha = f(\mathbb{R}^d)$ and $A_a = \{f = a\}$. Then $\{A_a : a \in \alpha\}$ is a partition of $\mathbb{R}^d$ and

$$\begin{aligned} E\|X - f(X)\|^r &= \sum_{a \in \alpha} \int_{A_a} \|x - a\|^r \, dP(x) \\ &\geq \sum_{a \in \alpha} V_r(P(\cdot|A_a))P(A_a). \end{aligned}$$

Conversely, for a Borel measurable partition $\mathcal{A}$ of $\mathbb{R}^d$ with $|\mathcal{A}| \leq n$, choose $a_A \in C_r(P \cdot |A)), A \in \mathcal{A}$, which is possible by Lemma 2.2 and let $f = \sum_{A \in \mathcal{A}} a_A 1_A$. (If $P(A) = 0$, let a_A be an arbitrary point in $\mathbb{R}^d$.) Then $f \in \mathcal{F}_n$ and

$$\sum_{A \in \mathcal{A}} V_r(P(\cdot|A))P(A) = \sum_{A \in \mathcal{A}} \int_A \|x - a_A\|^r \, dP(x) = E\|X - f(X)\|^r.$$

□

A Borel measurable partition $\mathcal{A}$ of $\mathbb{R}^d$ with $|\mathcal{A}| \leq n$ is called **n-optimal partition for P of order r** if

$$V_{n,r}(P) = \sum_{A \in \mathcal{A}} V_r(P(\cdot|A))P(A).$$

The proof of the preceding lemma shows that if f is an n-optimal quantizer, then $\{\{f = a\} : a \in f(\mathbb{R}^d)\}$ is an n-optimal partition. Conversely, if $\mathcal{A}$ is an n-optimal partition and $a_A \in C_r(P(\cdot|A))$ for $A \in \mathcal{A}$, then $f = \sum_{A \in \mathcal{A}} a_A 1_A$ is an n-optimal quantizer.

The quantization problem for P is further equivalent to the problem of approximating P by a discrete probability with at most n supporting points. For Borel probability measures P_1, P_2 on $\mathbb{R}^d$ with $\int \|x\|^r \, dP_i(x) < \infty$, let

$$\rho_r(P_1, P_2) = \inf_{\mu} \left(\int \|x - y\|^r \, d\mu(x, y) \right)^{1/r}, \tag{3.3}$$

where the infimum is taken over all Borel probabilities μ on $\mathbb{R}^d \times \mathbb{R}^d$ with fixed marginals P_1 and P_2. The **L_r-minimal metric ρ_r** (**L_r-Wasserstein metric** or **L_r-Kantorovich metric**) is appropriate for the quantization problem. This has been observed by Gray et al. (1975), Gray and Davisson (1975) and Pollard (1982a). By $\mathcal{P}_n$ denote the set of all discrete probabilities Q on $\mathbb{R}^d$ with $|\operatorname{supp}(Q)| \le n$.

3.4 Lemma

$$V_{n,r}(P) = \inf_{f \in \mathcal{F}_n} \rho_r^r(P, P^f) = \inf_{Q \in \mathcal{P}_n} \rho_r^r(P, Q),$$

where P^f denotes the image measure of P under f.

Proof

Given $f \in \mathcal{F}_n$, let μ_f denote the image measure of P under the map $\mathbb{R}^d \to \mathbb{R}^d \times \mathbb{R}^d$, $x \mapsto (x, f(x))$. Then

$$E\|X - f(X)\|^r = \int \|x - y\|^r \, d\mu_f(x, y) \ge \rho_r^r(P, P^f).$$

This implies

$$V_{n,r}(P) \ge \inf_{f \in \mathcal{F}_n} \rho_r^r(P, P^f) \ge \inf_{Q \in \mathcal{P}_n} \rho_r^r(P, Q).$$

If $Q \in \mathcal{P}_n$ with $Q(\alpha) = 1$, $|\alpha| \le n$, then for every Borel probability μ on $\mathbb{R}^d \times \mathbb{R}^d$ with marginals P and Q

$$\begin{aligned} \int \|x - y\|^r \, d\mu(x, y) &= \int_{\mathbb{R}^d \times \alpha} \|x - y\|^r \, d\mu(x, y) \\ &\ge \int_{\mathbb{R}^d \times \alpha} \min_{a \in \alpha} \|x - a\|^r \, d\mu(x, y) \\ &= \int_{\mathbb{R}^d} \min_{a \in \alpha} \|x - a\|^r \, dP(x), \end{aligned}$$

hence

$$\rho_r^r(P, Q) \ge E \min_{a \in \alpha} \|X - a\|^r.$$

By Lemma 3.1, this yields

$$\inf_{Q \in \mathcal{P}_n} \rho_r^r(P, Q) \geq V_{n,r}(P).$$

□

A measure $Q \in \mathcal{P}_n$ is called **n-optimal quantizing measure for P of order r** if

$$V_{n,r}(P) = \rho_r^r(P, Q).$$

If $f \in \mathcal{F}_n$ is an n-optimal quantizer, then $P^f \in \mathcal{P}_n$ is an n-optimal quantizing measure. Conversely, if $Q \in \mathcal{P}_n$ is an n-optimal quantizing measure and $\{A_a : a \in \alpha\}$ is a Voronoi partition with respect to $\alpha = \text{supp}(Q)$, then $f = \sum_{a \in \alpha} a 1_{A_a}$ is an n-optimal quantizer.

Several functional descriptions of ρ_r^r are known. Among them the most famous is the Kantorovich representation for $r = 1$

$$\rho_1(P_1, P_2) = \sup_g \left| \int g dP_1 - \int g dP_2 \right|,$$

where the supremum is taken over all functions $g\colon \mathbb{R}^d \to \mathbb{R}$ satisfying the Lipschitz condition $|g(x) - g(y)| \leq \|x - y\|$ for all $x, y \in \mathbb{R}^d$ In case $d = 1$, ρ_r admits the representation

$$\rho_r(P_1, P_2) = \left(\int_0^1 |F_1^{-1}(t) - F_2^{-1}(t)|^r dt \right)^{1/r}$$

and

$$\rho_1(P_1, P_2) = \int |F_1(t) - F_2(t)| dt,$$

where F_i denotes the distribution function and F_i^{-1} the quantile function of P_i ($F_i^{-1}(t) = \inf\{x \in \mathbb{R}\colon F_i(x) \geq t\}, t \in (0,1)$). For this background on L_r–minimal metrics we refer to Rachev (1991) and Rachev and Rüschendorf (1998, Chapters 2.5 and 2.6).

The empirical counterpart of quantization is cluster analysis. Somewhat more precisely, partitioning methods of cluster analysis for a finite sample according to a norm-based optimality criterion correspond to quantization for the empirical measure.

3.5 Example (Empirical version, cluster analysis)
Let $x_1, \ldots, x_k \in \mathbb{R}^d$ with $x_i = (x_{i1}, \ldots, x_{id})$ and let $P = \frac{1}{k} \sum_{i=1}^k \delta_{x_i}$ denote the empirical measure. We obtain from Lemma 3.3

$$V_{n,r}(P) = \frac{1}{k} \min_{\mathcal{C}} \sum_{C \in \mathcal{C}} \min_{a \in \mathbb{R}^d} \sum_{i \in C} \|x_i - a\|^r,$$

where the infimum is taken over all partitions $\mathcal{C}$ of $\{1, \ldots, k\}$ with $|\mathcal{C}| \leq n$. If the underlying norm is the l_2-norm, then

$$V_{n,2}(P) = \frac{1}{k} \min_{\mathcal{C}} \sum_{C \in \mathcal{C}} \sum_{i \in C} \|x_i - \overline{x}(C)\|^2,$$

where $\overline{x}(C) = \frac{1}{|C|} \sum_{i \in C} X_i$. This is the variance criterion for optimal grouping of data $x_1, \ldots, x_k$. If the underlying norm is the l_1-norm, then

$$V_{n,1}(P) = \frac{1}{k} \min_{\mathcal{C}} \sum_{C \in \mathcal{C}} \sum_{i \in C} \|x_i - \text{med}(C)\|,$$

where med(C) is an arbitrary element of $\mathsf{X}_{j=1}^d \text{med}(x_{ij}, i \in C)$ and med$(x_{ij}, i \in C)$ is the set of empirical medians of the real data $x_{ij}, i \in C$ (cf. Example 2.3 (b)). This is the l_1-criterion for optimal grouping of data. For treatments of cluster analysis which contain discussions of the above optimality criteria we refer to Bock (1974) and Späth (1985).

The n-optimal sets of centers for P of order r correspond to global minima of the function

$$\psi_{n,r} : (\mathbb{R}^d)^n \to \mathbb{R}_+, \; \psi_{n,r}(a_1, \ldots, a_n) = E \min_{1 \leq i \leq n} \|X - a_i\|^r. \tag{3.4}$$

Notice that $\psi_{1,r} = \psi_r$. While ψ_r is convex, $\psi_{n,r}$ is typically not convex for $n \geq 2$. Therefore, local minimum points of $\psi_{n,r}$ may not be global minimum points of $\psi_{n,r}$. The lack of any straightforward solution for the quantization problem (at least for $d \geq 2$) is a result of the difficulty in dealing with the nonconvex nature of quantization.

Notes

Treatments of the quantization problem with applications in information theory (analog-to-digital conversion, signal compression, coding theory) are contained in the March 1982 Special Issue of IEEE Transactions on Information Theory (Vol. 28, pp. 127-202), in Gray (1990), Abut (1990), Gersho and Gray (1992), and in Calderbank et al. (1993). Some material may also be found in Fang and Wang (1994). The n-th quantization error $V_{n,r}(P)$ appears in error bounds for numerical integration; see Pagès (1997). In the one-dimensional case the quantization problem for $r = 2$ corresponds to the optimal stratification problem for Bowley (or proportional) sampling schemes of Dalenius (1950). Also in the one-dimensional case the quantization problem can be seen as optimal knot selection for piecewise constant L_r-approximation. A review of the problem in this spirit for $r = 2$ can be found in Eubank (1988). A fuzzy version of the quantization problem is discussed by Yang and Yu (1991).

Let us mention that n-optimal sets of centers are sometimes called sets of principal points or representative points.

Occasionally, it may be preferable to use other measures for the quantization error than L_r-metrics as in (3.1). The limiting case of the L_∞-metric ("worst-case error")

$$\text{ess sup}\|X - f(X)\| = \inf\{c \geq 0\colon \mathbb{P}(\|X - f(X)\| > c) = 0\}, f \in \mathcal{F}_n$$

is studied in Section 10 and the Ky Fan metric

$$\inf\{\varepsilon \geq 0\colon \mathbb{P}(\|X - f(X)\| > \varepsilon) \leq \varepsilon\}$$

is studied in Graf and Luschgy (1999a). While the first metric requires X to be bounded, the latter does not. The Ky Fan error measure leads to the approximation problem for P with respect to the Prohorov metric (in the sense of Lemma 3.4). An investigation of the quantization problem based on the geometric mean error

$$\exp E \log \|X - f(X)\|$$

as measure of performance can be found in Graf and Luschgy (1999b). Input weighted error measures of the form

$$E(X - f(X))^t B(X)(X - f(X)),$$

where $B(x)$ is a positive definite matrix for every $x \in \mathbb{R}^d$, have proved useful in speech and image compression systems. For various aspects of the quantization problem based on this error see e.g. Gray and Karnin (1982), Gardner and Rao (1995), Li et al. (1999) and Linder et al. (1999).

Basically different quantization problems have been treated by Elias (1970) and more recently by Bock (1992) and Pötzelberger and Strasser (1999).

4 Basic properties of optimal quantizers

As in the previous section, let X be a $\mathbb{R}^d$-valued random variable with distribution P such that $E\|X\|^r < \infty$ for some $1 \leq r < \infty$. Further, we assume (with the only exception of the last subsection) $n \geq 2$ and in order to avoid trivial cases, we also assume $P \notin \mathcal{P}_{n-1}$, that is, $|\operatorname{supp}(P)| \geq n$.

4.1 Stationarity and existence

The following two theorems provide necessary conditions for n-optimality of quantizers. They provide the gateway to most available algorithmic solutions.

4.1 Theorem (Necessary conditions for optimality)
Let $\alpha \in C_{n,r}(P)$ and let $\{A_a : a \in \alpha\}$ be a Voronoi partition of $\mathbb{R}^d$ with respect to α and P. Then

$$|\alpha| = n, \; P(A_a) > 0 \text{ for every } a \in \alpha,$$
$$\beta \in C_{m,r}(P(\cdot \mid \bigcup_{a \in \beta} A_a)) \text{ for every } \beta \subset \alpha \text{ with } |\beta| = m.$$

In particular,

$$P(W(a|\alpha)) > 0, \; a \in C_r(P(\cdot|W(a|\alpha))) \text{ for every } a \in \alpha. \tag{4.1}$$

Proof

Let $\gamma = \{a \in \alpha : P(A_a) > 0\}$ and assume $|\gamma| < n$. Obviously, $\gamma \in C_{n,r}(P)$. Since $P \notin \mathcal{P}_{n-1}$, there exists $a \in \gamma$ such that $P(\cdot|A_a)$ is not a point mass. We can conclude that

$$P(H(a,b)^c \cap A_a) > 0$$

for some $b \in \mathbb{R}^d$. (Recall $H(a,b) = \{x \in \mathbb{R}^d : \|x-a\| \leq \|x-b\|\}$.) In fact, we have $P(A_a \setminus \{a\}) > 0$ and hence, there is a compact set $K \subset A_a \setminus \{a\}$ with $P(K) > 0$. Since $K \subset \bigcup_{b \in K} H(a,b)^c$ and $H(a,b)^c$ is open, we can find a finite subset B of K such that $K \subset \bigcup_{b \in B} H(a,b)^c$. This gives the existence of a point $b \in B$ with the required property. It follows that

$$V_{n,r}(P) = E \min_{a \in \gamma} \|X-a\|^r > E \min_{a \in \gamma \cup \{b\}} \|X-a\|^r \geq V_{n,r}(P),$$

a contradiction.

As for the assertion concerning β, assume $\beta \notin C_{m,r}(P(\cdot| \bigcup_{a \in \beta} A_a))$. Then there exists $\delta \subset \mathbb{R}^d$ with $|\delta| \leq m$ and

$$\int_{\bigcup_{a \in \beta} A_a} \min_{b \in \beta} \|x-b\|^r \, dP(x) > \int_{\bigcup_{a \in \beta} A_a} \min_{b \in \delta} \|x-b\|^r \, dP(x).$$

It follows that

$$V_{n,r}(P) = E \min_{a \in \alpha} \|X - a\|^r > E \min_{a \in \delta \cup (\alpha \setminus \beta)} \|X - a\|^r \geq V_{n,r}(P),$$

a contradiction. □

We know from (1.8) and Theorem 1.5 that the Voronoi diagram of every finite subset of $\mathbb{R}^d$ is a P-tesselation provided the underlying norm is strictly convex and P is absolutely continuous with respect to λ^d. So the following result is of interest for probability distributions P which are not absolutely continuous with respect to λ^d. (Such probabilities are considered in Chapter III.)

4.2 Theorem (Necessary condition for optimality)
Let $\alpha \in C_{n,r}(P)$ and let $r > 1$ or $P(\alpha) = 0$. Suppose the underlying norm is strictly convex and smooth. Then the Voronoi diagram of α is a P-tesselation of $\mathbb{R}^d$.

Proof

We have to prove

$$P(W(a|\alpha) \cap W(b|\alpha)) = 0$$

for every $a, b \in \alpha, a \neq b$. Fix $a, b \in \alpha, a \neq b$ and assume $P(W(a|\alpha) \cap W(b|\alpha)) > 0$. Choose a Voronoi partition $\{A_c : c \in \alpha\}$ with respect to α such that $A_a = W(a|\alpha) \setminus W(b|\alpha)$. Then by Theorem 4.1,

$$a \in C_r(P(\cdot|A_a)) \cap C_r(P(\cdot|W(a|\alpha))).$$

From Lemma 2.5 it follows that

$$\int_{A_a \setminus \{a\}} \|x - a\|^{r-1} \nabla \| \, \|(a - x) \, dP(x) = 0$$

and

$$\int_{W(a|\alpha) \setminus \{a\}} \|x - a\|^{r-1} \nabla \| \, \|(a - x) \, dP(x) = 0$$

which yields

$$\int_{W(a|\alpha) \cap W(b|\alpha)} \|x - a\|^{r-1} \nabla \| \, \|(a - x) \, dP(x) = 0.$$

Therefore, again by Lemma 2.5, $a \in C_r(Q)$ with $Q = P(\cdot|W(a|\alpha) \cap W(b|\alpha))$. Since $W(a|\alpha) \cap W(b|\alpha)$ is contained in the separator $S(a, b)$, this implies $b \in C_r(Q)$. Thus,

Q has two different centers a and b of order r. By Theorem 2.4, this can happen only if $r = 1$ and $Q(L) = 1$, where L is the line through a and b. Since

$$L \cap W(a|\alpha) \cap W(b|\alpha) \subset L \cap S(a,b) = \{(a+b)/2\},$$

one obtains $Q = \delta_{(a+b)/2}$. It follows

$$\{a,b\} \subset C_1(Q) = \{(a+b)/2\},$$

a contradiction. □

A set $\alpha \subset \mathbb{R}^d$ with $|\alpha| = n$ satisfying condition (4.1) is called **n-stationary set of centers for P of order r**. Let $S_{n,r}(P)$ denote the set of all these n-stationary sets for P and denote by $SS_{n,r}(P)$ the subset of $S_{n,r}(P)$ consisting of all $\alpha \in S_{n,r}(P)$ such that the Voronoi diagram of α is a P-tesselation. Then by Theorem 4.1,

$$C_{n,r}(P) \subset S_{n,r}(P).$$

Note that any Voronoi partition $\{A_a : a \in \alpha\}$ with respect to $\alpha \in SS_{n,r}(P)$ and P satisfies $A_a = W(a|\alpha)$ P-a.s., $a \in \alpha$. We also write $S_{n,r}(X)$ and $SS_{n,r}(X)$ instead of $S_{n,r}(P)$ and $SS_{n,r}(P)$, respectively.

4.3 Corollary

(a) *Let $\mathcal{A}$ be an n-optimal partition for P of order r. Then $|\mathcal{A}| = n, P(A) > 0$ for every $A \in \mathcal{A}, C_r(P \cdot |A)) \cap C_r(P(\cdot|B)) = \emptyset$ for every $A, B \in \mathcal{A}, A \neq B$, and $\mathcal{A}$ is a Voronoi partition of $\mathbb{R}^d$ with respect to $\alpha = \{a_A : A \in \mathcal{A}\}$ and P for any choice of $a_A \in C_r(P(\cdot|A))$.*

(b) *Let $f \in \mathcal{F}_n$ be an n-optimal quantizer for P of order r and let $\alpha = f(\mathbb{R}^d)$. Then $\alpha \in S_{n,r}(P), \{\{f = a\} : a \in \alpha\}$ is a Voronoi partition of $\mathbb{R}^d$ with respect to α and P, $P(\{f = a\}) > 0$ and $a \in C_r(P(\cdot|\{f = a\}))$ for every $a \in \alpha$.*

Proof

(a) We have

$$\begin{aligned} V_{n,r}(P) &= \sum_{A \in \mathcal{A}} V_r(P(\cdot|A))P(A) = \sum_{A \in \mathcal{A}} \int_A \|x - a_A\|^r \, dP(x) \\ &\geq \sum_{A \in \mathcal{A}} \int_A \min_{b \in \alpha} \|x - b\|^r \, dP(x) = \int \min_{b \in \alpha} \|x - b\|^r \, dP(x) \\ &\geq V_{n,r}(P). \end{aligned}$$

This implies $\alpha \in C_{n,r}(P)$ and

$$\int_A \|x - a_A\|^r \, dP(x) = \int_A \min_{b \in \alpha} \|x - b\|^r \, dP(x), \; A \in \mathcal{A}.$$

Therefore, $\mathcal{A}$ is a Voronoi partition of $\mathbb{R}^d$ with respect to α and P. The remaining assertions follow from Theorem 4.1.

(b) As in (a) one can check that $\{\{f = a\} : a \in \alpha\}$ is a Voronoi partition of $\mathbb{R}^d$ with respect to α and P. The remaining assertions follow from Theorem 4.1. □

Under the condition $C_{n,r}(P) \subset SS_{n,r}(P)$ there is a characterization of optimal quantizing measures.

4.4 Lemma
Suppose $C_{n,r}(P) \subset SS_{n,r}(P)$, that is, the Voronoi diagram of every $\alpha \in C_{n,r}(P)$ is a P-tesselation of $\mathbb{R}^d$. Then the set of n-optimal quantizing measures for P of order r coincides with the set

$$\{P^f : f \in \mathcal{F}_n \text{ } n\text{-optimal for } P \text{ of order } r\}.$$

Proof

Let $Q = \sum_{a\in\alpha} p_a \delta_a$ be an n-optimal quantizing measure of order r. Choose a Borel probability μ on $\mathbb{R}^d \times \mathbb{R}^d$ with marginals P and Q such that $\rho_r(P,Q)^r = \int \|x - y\|^r \, d\mu(x,y)$ and let $f = \sum_{a\in\alpha} a 1_{A_a}$, where $\{A_a : a \in \alpha\}$ denotes a Voronoi partition of $\mathbb{R}^d$ with respect to α. Then f is an n-optimal quantizer and $\alpha \in C_{n,r}(P)$. Therefore

$$\begin{aligned}
\sum_{a\in\alpha} \int_{\mathbb{R}^d\times\{a\}} \|x-a\|^r \, d\mu(x,y) &= \int_{\mathbb{R}^d\times\alpha} \|x-y\|^r \, d\mu(x,y) \\
&= \rho_r(P,Q)^r = V_{n,r}(P) \\
&= \int \min_{b\in\alpha} \|x-b\|^r \, dP(x) \\
&= \sum_{a\in\alpha} \int_{\mathbb{R}^d\times\{a\}} \min_{b\in\alpha} \|x-b\|^r \, d\mu(x,y).
\end{aligned}$$

This implies

$$\mathbb{R}^d \times \{a\} \subset W(a|\alpha) \times \mathbb{R}^d \quad \mu\text{-a.s.}$$

for every $a \in \alpha$. Hence $p_a \le P(W(a|\alpha)), a \in \alpha$. It follows from the assumption that $\sum_{a\in\alpha} P(W(a|\alpha)) = 1$. Since $\sum_{a\in\alpha} p_a = \sum_{a\in\alpha} P(A_a) = 1$, one obtains $p_a = P(W(a|\alpha)) = P(A_a), a \in \alpha$. This gives $Q = P^f$. The converse inclusion was already mentioned in Section 3. □

In general, the Voronoi diagram of an n-optimal set of centers $\alpha \in C_{n,r}(P)$ need not be a P-tesselation and also, the assertion of Lemma 4.4 may fail. This is exhibited by the following example.

4.5 Example
Let the underlying norm on $\mathbb{R}^2$ be the l_∞-norm. Consider $P = \frac{1}{4}(\delta_{(-1,0)} + \delta_{(0,1)} + \delta_{(1,0)} + \delta_{(0,-1)})$ and let $n = 2, r = 1$. It is geometrically rather obvious that $V_{2,1}(P) = 1/2$

and $C_{2,1}(P)$ consists of all sets $\{a, b\}$ with $a, b \in \{x \in \mathbb{R}^2 : |x_1| + |x_2| = 1\}$ such that the line segment joining a and b meets the line $\{x_1 = 0\}$; see Figure 4.1.

Now let $a = (-1, 0)$ and $b = (1, 0)$. Then $\{a, b\} \in C_{2,1}(P)$ and $S(a, b)$ contains the line through $(0, -1)$ and $(0, 1)$. One obtains $P(S(a, b)) = 1/2 > 0$ and hence, the Voronoi diagram $\{H(a, b), H(b, a)\}$ of $\{a, b\}$ is not a P-tesselation. Furthermore, the probability $Q = \frac{5}{8}\delta_a + \frac{3}{8}\delta_b$ is a 2-optimal quantizing measure for P. In fact, let $x_1 = (-1, 0), x_2 = (0, 1), x_3 = (1, 0), x_4 = (0, -1)$, and define a discrete probability μ on $\mathbb{R}^2 \times \mathbb{R}^2$ by

$$\mu(\{(x_1, a)\}) = \mu(\{(x_3, b)\}) = \mu(\{(x_4, a)\}) = 1/4,$$
$$\mu(\{(x_2, a)\}) = \mu(\{(x_2, b)\}) = 1/8.$$

Then the marginals of μ are P and Q, respectively, and

$$\int \|x - y\| \, d\mu(x, y) = 1/2.$$

This yields $V_{2,1}(P) = \rho_1(P, Q)$. Obviously, $Q \neq P^f$ for every $f \in \mathcal{F}_2$.

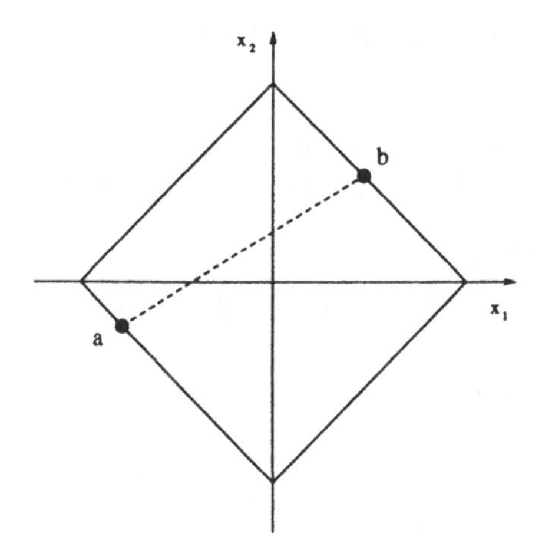

Figure 4.1: 2-optimal centers of order 1 with respect to the l_∞ norm

4.6 Remark (Euclidean norms)
Let $\|x\| = \langle x, x \rangle^{1/2}$ for some scalar product $\langle \, , \, \rangle$ on $\mathbb{R}^d$.

(a) We have

$$\bigcup \{\alpha : \alpha \in S_{n,r}(P)\} \subset \operatorname{cl} \operatorname{conv}(\operatorname{supp}(P)).$$

This follows from Lemma 2.6(a). Furthermore, by Theorem 4.2, $C_{n,r}(P) \subset SS_{n,r}(P)$ provided $r > 1$. Recall that in case $r = 2$, the second condition of (4.1) means $a = E(X | X \in W(a|\alpha)), a \in \alpha$.

(b) If $\alpha \in SS_{n,2}(P)$, then

$$\sum_{a \in \alpha} a P(W(a|\alpha)) = EX.$$

In case $d = 1$, a simple but sometimes useful consequence is that $\min \alpha < EX < \max \alpha$ holds for $\alpha \in SS_{n,2}(P)$ $(n \geq 2)$.

(c) If $\alpha \in SS_{n,2}(P)$, then

$$\begin{aligned} E \min_{a \in \alpha} \|X - a\|^2 &= E\|X\|^2 - \sum_{a \in \alpha} \|a\|^2 P(W(a|\alpha)) \\ &= V_2(X) + \|EX\|^2 - \sum_{a \in \alpha} \|a\|^2 P(W(a|\alpha)). \end{aligned}$$

Hence, for $\alpha \subset \mathbb{R}^d$ we have $\alpha \in C_{n,2}(P)$ if and only if $\alpha \in SS_{n,2}(P)$ and

$$\sum_{a \in \alpha} \|a\|^2 P(W(a|\alpha)) = \max_{\beta \in SS_{n,r}(P)} \sum_{b \in \beta} \|b\|^2 P(W(b|\beta)).$$

(d) If $f \in \mathcal{F}_n$ is an n-optimal quantizer for P of order 2, then

$$\begin{aligned} &Ef(X) = EX, \\ &E\langle X - f(X), f(X)\rangle = 0, \\ &V_{n,2}(X) = E\|X - f(X)\|^2 = E\|X\|^2 - E\|f(X)\|^2. \end{aligned}$$

This follows from (b) and (c).

The sets $S_{n,r}(X)$ and $SS_{n,r}(X)$ have the same equivariance property as $C_{n,r}(X)$.

4.7 Lemma
Let $T : \mathbb{R}^d \to \mathbb{R}^d$ be a similarity transformation. Then

$$\begin{aligned} S_{n,r}(T(X)) &= TS_{n,r}(X), \\ SS_{n,r}(T(X)) &= TSS_{n,r}(X). \end{aligned}$$

Proof

Easy consequence of the equivariance properties of Voronoi regions and $C_r(X)$ given in Lemmas 1.6 and 2.1 (a). □

Stationary product quantizers are discussed in the following lemma.

4.8 Lemma (Product quantizers)
Let the underlying norm be the l_r-norm. Let $n_i \in \mathbb{N}$, $\beta_i \subset \mathbb{R}$ with $|\beta_i| \leq n_i$, $1 \leq i \leq d$, and $\alpha = \times_{i=1}^d \beta_i$.

(a) *Suppose that $X_1, \ldots, X_d$ are independent and let $n = \prod_{i=1}^d n_i$. Then $\alpha \in S_{n,r}(X)$ if and only if $\beta_i \in S_{n_i,r}(X_i)$ for every i.*

(b) *If* $\beta_i \in C_{n_i,r}(X_i)$ *for every* i, *then*

$$E \min_{a\in\alpha} \|X - a\|^r = \sum_{i=1}^{d} V_{n_i,r}(X_i).$$

Proof

(a) Let P_i denote the distribution of $X_i, 1 \leq i \leq d$. For $a = (b_1, \dots, b_d) \in \alpha$ with $b_i \in \beta_i$ for every i, we have

$$W(a|\alpha) = \bigtimes_{i=1}^{d} W(b_i|\beta_i).$$

Assume $\alpha \in S_{n,r}(X)$. Then $\prod_{i=1}^{d} |\beta_i| = |\alpha| = n$ and

$$\prod_{i=1}^{d} P_i(W(b_i|\beta_i)) = P(W(a|\alpha)) > 0, \ a = (b_1, \dots, b_d) \in \alpha$$

which gives $|\beta_i| = n_i$ and $P_i(W(b_i|\beta_i)) > 0$ for every i. Fix i and let $c = (c_1, \dots, c_d) \in \mathbb{R}^d$ with $c_j = b_j$ for $j \neq i$. Then

$$\begin{aligned}\sum_{j=1}^{d} \int_{W(a|\alpha)} |x_j - b_j|^r \, dP(x) &= \int_{W(a|\alpha)} \|x - a\|^r \, dP(x) \\ &\leq \int_{W(a|\alpha)} \|x - c\|^r \, dP(x) \\ &= \sum_{j\neq i} \int_{W(a|\alpha)} |x_j - b_j|^r \, dP(x) + \int_{W(a|\alpha)} |x_i - c_i|^r \, dP(x)\end{aligned}$$

and hence

$$\int_{W(a|\alpha)} |x_i - b_i|^r \, dP(x) \leq \int_{W(a|\alpha)} |x_i - c_i|^r \, dP(x).$$

Since for $c_i \in \mathbb{R}$

$$\int_{W(b_i|\beta_i)} |x_i - c_i|^r \, dP_i(x_i)/P_i(W(b_i|\beta_i)) = \int_{W(a|\alpha)} |x_i - c_i|^r \, dP(x)/P(W(a|\alpha)),$$

this yields $b_i \in C_r(P_i(\cdot|W(b_i|\beta_i)))$. Therefore, $\beta_i \in S_{n_i,r}(X_i)$.

Conversely, assume $\beta_i \in S_{n_i,r}(X_i)$ for every i. Then $|\alpha| = n$ and $P(W(a|\alpha)) > 0$. Let $c = (c_1, \ldots, c_d) \in \mathbb{R}^d$. One obtains

$$\begin{aligned}\int_{W(a|\alpha)} \|x-a\|^r \, dP(x)/P(W(a|\alpha)) &= \sum_{i=1}^d \int_{W(b_i|\beta_i)} |x_i - b_i|^r \, dP_i(x_i)/P_i(W(b_i|\beta_i)) \\ &\leq \sum_{i=1}^d \int_{W(b_i|\beta_i)} |x_i - c_i|^r \, dP_i(x_i)/P_i(W(b_i|\beta_i)) \\ &= \int_{W(a|\alpha)} \|x-c\|^r \, dP(x)\end{aligned}$$

and hence $a \in C_r(P(\cdot|W(a|\alpha)))$. Therefore, $\alpha \in S_{n,r}(X)$.

(b) We have

$$\begin{aligned}E \min_{a\in\alpha} \|X-a\|^r &= \sum_{i=1}^d E \min_{b\in\beta_i} |X_i - b|^r \\ &= \sum_{i=1}^d V_{n_i,r}(X_i).\end{aligned}$$

□

The n-stationary sets for P are related to the stationary points of the function $\psi_{n,r}$ (see (3.4)).

4.9 Lemma
$\psi_{n,r}$ is continuous on $(\mathbb{R}^d)^n$.

Proof

Immediate consequence of the continuity of $(a_1, \ldots, a_n) \mapsto \min_{1\leq i\leq n} \|x - a_i\|^r$ for every $x \in \mathbb{R}^d$ and Lebesgue's dominated convergence theorem. □

4.10 Lemma
Let $a_1, \ldots, a_n \in \mathbb{R}^d$ with $a_i \neq a_j$ for $i \neq j$. Suppose the Voronoi diagram of $\alpha = \{a_1, \ldots, a_n\}$ is a P-tesselation of $\mathbb{R}^d$. Then $\psi_{n,r}$ has a one-sided directional derivate at $a = (a_1, \ldots, a_n)$ in every direction $y = (y_1, \ldots, y_n) \in (\mathbb{R}^d)^n$ given by

$$\nabla_+ \psi_{n,r}(a, y) = r \sum_{i=1}^n \int_{W(a_i|\alpha)} \|x - a_i\|^{r-1} \nabla_+ \|\ \|(a_i - x, y_i) \, dP(x).$$

If the underlying norm is smooth and furthermore, $r > 1$ or $P(\alpha) = 0$, then $\psi_{n,r}$ is differentiable at the point a with derivative

$$\nabla \psi_{n,r}(a) = \Big(r \int_{W(a_i|\alpha)\setminus\{a_i\}} \|x - a_i\|^{r-1} \nabla \|\ \|(a_i - x) \, dP(x) \Big)_{1\leq i\leq n}.$$

Proof

Recall that

$$W_0(a_i|\alpha) = \bigcap_{j\neq i} \{x \in \mathbb{R}^d : \|x - a_i\| < \|x - a_j\|\}.$$

For $b = (b_1, \ldots, b_n) \in (\mathbb{R}^d)^n$ set $d(x,b) = \min_{1\leq i\leq n} \|x - b_i\|$. By assumption, the Voronoi diagram of α is a P-tesselation of $\mathbb{R}^d$ which gives $P\big(\bigcup_{i=1}^n W_0(a_i|\alpha)\big) = 1$. Furthermore, we have

$$\begin{aligned} |d(x, a+b) - d(x,a)| &\leq \max_{1\leq i\leq n} |\, \|x - a_i - b_i\| - \|x - a_i\| \,| \\ &\leq \max_{1\leq i\leq n} \|b_i\| \end{aligned}$$

and since

$$|u^r - v^r| \leq r \max\{u^{r-1}, v^{r-1}\}|u - v|, \quad u, v \geq 0,$$

we obtain

$$|d(x, a+b)^r - d(x,a)^r| \leq (C_1\|x\|^{r-1} + C_2) \max_{1\leq i\leq n} \|b_i\|, \tag{4.2}$$

$x \in \mathbb{R}^d, b \in (\mathbb{R}^d)^n$ with $\max_{1\leq i\leq n} \|b_i\| \leq 1$ for numerical constants $C_1, C_2 \geq 0$ not depending on x and b.

Let $y = (y_1, \ldots, y_n) \in (\mathbb{R}^d)^n$. Then

$$t^{-1}(\psi_{n,r}(a + ty) - \psi_{n,r}(a)) = \sum_{i=1}^n \int_{W_0(a_i|\alpha)} t^{-1}(d(x, a+ty)^r - d(x,a)^r)\, dP(x).$$

For $x \in W_0(a_i|\alpha)$, there exists $\varepsilon > 0$ such that the $\mathbb{R}^d$-components of $a + ty$ are pairwise different and

$$x \in W(a_i + ty_i | \{a_1 + ty_1, \ldots, a_n + ty_n\})$$

for every $0 < t \leq \varepsilon$. This implies

$$\begin{aligned} t^{-1}(d(x, a+ty)^r - d(x,a)^r) &= t^{-1}(\|x - (a_i + ty_i)\|^r - \|x - a_i\|^r) \\ &\to \nabla_+ \|\ \|^r(a_i - x, y_i) \quad \text{as } t \to 0+. \end{aligned}$$

Thus the assertion about the one-sided directional derivative of $\psi_{n,r}$ follows from Lebesgue's dominated convergence theorem in view of (4.2).

Now assume that the underlying norm is smooth. Then we have

$$(\max_{1\le j\le n}\|b_j\|)^{-1}\Big(\psi_{n,r}(a+b)-\psi_{n,r}(a)-\sum_{i=1}^{n}\Big\langle\int\limits_{W(a_i|\alpha)\setminus\{a_i\}}\nabla\|\ \|^r(a_i-x)\,dP(x),b_i\Big\rangle\Big)$$

$$=(\max_{1\le j\le n}\|b_j\|)^{-1}\sum_{i=1}^{n}\int\limits_{W_0(a_i|\alpha)}(d(x,a+b)^r-d(x,a)^r-\langle\nabla\|\ \|^r(a_i-x),b_i\rangle)\,dP(x)$$

where $\langle x,z\rangle=\sum_{j=1}^{d}x_jz_j$, $x,z\in\mathbb{R}^d$. For $x\in W_0(a_i|\alpha)$, there exists $\varepsilon>0$ such that the $\mathbb{R}^d$-components of $a+b$ are pairwise different and

$$x\in W(a_i+b_i|\{a_1+b_1,\dots,a_n+b_n\})$$

for every $b\in(\mathbb{R}^d)^n$ with $\max_{1\le j\le n}\|b_j\|\le\varepsilon$. This implies

$$(\max_{1\le j\le n}\|b_j\|)^{-1}(d(x,a+b)^r-d(x,a)^r-\langle\nabla\|\ \|^r(a_i-x),b_i\rangle)\to 0$$
$$\text{as }\max_{1\le j\le n}\|b_j\|\to 0\quad(x\ne a_i,\ \text{if } r=1).$$

In view of (4.2), the assertion about $\nabla\psi_{n,r}(a)$ follows from Lebesgues's dominated convergence theorem. □

Consequently, in view of Lemma 2.5, n-stationary sets $\alpha\in SS_{n,r}(P)$ of centers provide stationary points of $\psi_{n,r}$, i.e. $\nabla_+\psi_{n,r}(a,y)\ge 0$ for every $y\in(\mathbb{R}^d)^n$. The following example taken from Lloyd (1982) shows that a n-stationary set of centers does not necessarily yield a local minimum point of $\psi_{n,r}$.

4.11 Example
Let $P=c_1U([-1,0])+c_2U([0,1])$ with $c_2>c_1>0, c_1+c_2=1$, and let $n=2, r=2$. Then $EX=\frac{1}{2}(c_2-c_1), EX^2=\frac{1}{3}$ and $\{-\frac{1}{2},\frac{1}{2}\}\in S_{2,2}(P)$. For $-1\le a_1\le a_2\le 1$, we have

$$\psi(a_1,a_2)=\frac{c_1}{3}\left[\frac{(a_2-a_1)^3}{4}+(1+a_1)^3-a_2^3\right]+\frac{c_2}{3}\left[(1-a_2)^3+a_2^3\right]\ \text{ if } a_1+a_2\le 0,$$
$$\psi(a_1,a_2)=\frac{c_1}{3}\left[(1+a_1)^3-a_1^3\right]+\frac{c_2}{3}\left[\frac{(a_2-a_1)^3}{4}+a_1^3+(1-a_2)^3\right]\ \text{ if } a_1+a_2>0,$$

where $\psi=\psi_{2,2}$. One obtains $\psi\left(-\frac{1}{2},\frac{1}{2}\right)=\frac{1}{12}$ and

$$\psi\left(-\frac{1}{2}+\varepsilon,\frac{1}{2}\right)=\frac{1}{12}+c_1\varepsilon^2+c_2\left(\frac{\varepsilon^3}{4}-\frac{\varepsilon^2}{4}\right)<\frac{1}{12}\ \text{for every } 0<\varepsilon<5-\frac{4}{c_2}$$

provided $c_2>\frac{4}{5}$. Thus $(-\frac{1}{2},\frac{1}{2})$ is not a local minimum point of ψ in case $c_2>\frac{4}{5}$. (It is also not a local maximum point of ψ.) We have

$$S_{2,2}(P)=C_{2,2}(P)=\left\{\left\{-\frac{1}{2},\frac{1}{2}\right\}\right\}\ \text{ if } c_2\le\frac{3}{4},$$

$$C_{2,2}(P) = \left\{\left\{\frac{10c_2 - 9}{4c_2}, \frac{6c_2 - 3}{4c_2}\right\}\right\} \text{ and}$$
$$S_{2,2}(P) = C_{2,2}(P) \cup \left\{\left\{-\frac{1}{2}, \frac{1}{2}\right\}\right\} \text{ if } c_2 > \frac{3}{4}.$$

The next theorem ensures the existence of n-optimal quantizers. We follow the lines of Pollard's (1982a) proof for the euclidean case and $r = 2$.

4.12 Theorem (Existence)
We have $V_{n,r}(P) < V_{n-1,r}(P)$. The level set $\{\psi_{n,r} \le c\}$ is compact for every $0 \le c < V_{n-1,r}(P)$. In particular, $C_{n,r}(P)$ is not empty and $\bigcup\{\alpha : \alpha \in C_{n,r}(P)\}$ is a bounded subset of $\mathbb{R}^d$.

Proof

By Lemma 4.9, the level sets of $\psi_{n,r}$ are closed. Choose $0 < s < S$ (depending on n, r, P and c) such that

$$P(B(0,s)) > 0, \; (S-s)^r P(B(0,s)) > c,$$
$$2^r \int_{B(0,2S)^c} \|x\|^r \, dP(x) < V_{n-1,r}(P) - c.$$

Let $(a_1, \ldots, a_n) \in \{\psi_{n,r} \le c\}$. Since $c < V_{n-1,r}(P)$, we have $a_i \ne a_j$ for $i \ne j$. Assume without loss of generality $\|a_1\| \le \ldots \le \|a_n\|$. Then $\|a_1\| \le S$. Otherwise

$$c \ge \int_{B(0,s)} \min_{1 \le i \le n} \|x - a_i\|^r \, dP(x) \ge (S-s)^r P(B(0,s)),$$

a contradiction. Furthermore, $\|a_n\| \le 5S$. Otherwise

$$\|x - a_1\| \le \|x - a_n\| 1_{B(0,2S)}(x) + 2\|x\| 1_{B(0,2S)^c}(x)$$

for every $x \in \mathbb{R}^d$ because $\|a_1\| \le S$. Therefore, if $\{A_1, \ldots, A_n\}$ denotes a Voronoi partition of $\mathbb{R}^d$ with respect to $\{a_1, \ldots, a_n\}$, one gets

$$\begin{aligned}
V_{n-1,r}(P) &\le \psi_{n-1,r}(a_1, \ldots, a_{n-1}) \\
&= \sum_{j=1}^{n} \int_{A_j} \min_{1 \le i \le n-1} \|x - a_i\|^r \, dP(x) \\
&\le \sum_{j=1}^{n-1} \int_{A_j} \|x - a_j\|^r \, dP(x) + \int_{A_n} \|x - a_1\|^r \, dP(x) \\
&\le \sum_{j=1}^{n} \int_{A_j} \|x - a_j\|^r \, dP(x) + 2^r \int_{B(0,2S)^c} \|x\|^r \, dP(x) \\
&< \psi_{n,r}(a_1, \ldots, a_n) + V_{n-1,r}(P) - c \le V_{n-1,r}(P),
\end{aligned}$$

a contradiction. We thus obtain

$$\{\psi_{n,r} \le c\} \subset \mathop{\times}_{1}^{n} B(0, 5S).$$

Hence, if $V_{n,r}(P) < c < V_{n-1,r}(P)$, the level set $\{\psi_{n,r} \le c\}$ is not empty and compact. This implies that $\{\psi_{n,r} = V_{n,r}(P)\}$ is not empty and compact and, in particular, $C_{n,r}(P) \ne \emptyset$ provided $V_{n,r}(P) < V_{n-1,r}(P)$.

Finally, we observe that this condition holds. We have $C_r(P) \ne \emptyset$ by Lemma 2.2. Therefore, $V_{2,r}(P) < V_r(P)$, since otherwise there exists an 2-optimal set α of centers with $|\alpha| = 1$ which contradicts Theorem 4.1. Proceed inductively: if $V_{m,r}(P) < V_{m-1,r}(P)$ for some $2 \le m \le n-1$, then $C_{m,r}(P) \ne \emptyset$ by the preceding part of the proof and hence, $V_{m+1,r}(P) < V_{m,r}(P)$ again by Theorem 4.1. □

From Example 4.5 we know that there may be more than one n-optimal set of centers for P. Here is another example of this fact for an univariate symmetric distribution.

4.13 Example
Let P denote the uniform distribution on $[-2,-1] \cup [1,2]$ and let $n = 3, r = 2$. Then

$$C_{3,2}(P) = S_{3,2}(P) = \left\{ \left\{ \frac{-7}{4}, \frac{-5}{4}, \frac{3}{2} \right\}, \left\{ \frac{-3}{2}, \frac{5}{4}, \frac{7}{4} \right\} \right\}.$$

(Note that $\alpha = \{-\frac{3}{2}, 0, \frac{3}{2}\}$ is not a 3-stationary set, since $P(W(0|\alpha)) = P([-\frac{3}{4}, \frac{3}{4}]) = 0$. However, $(-\frac{3}{2}, 0, \frac{3}{2})$ is a stationary point of $\psi_{3,2}$.) Here we have $V_2(P) = \operatorname{Var} X = 7/3$ and $V_{3,2}(P) = 5/96$. Thus already 3-level quantization reduces the variance considerably.

For dimensions $d \ge 2$, typically $|C_{n,r}(P)| \ge 2$ holds. This is related to the equivariance property of $C_{n,r}(P)$ (see Lemma 3.2). A uniqueness criterion for univariate distributions is discussed in the next section.

4.2 The functional $V_{n,r}$

The following simple properties of the n-th quantization error functional turn out to be useful.

4.14 Lemma
Let $P = \sum_{i=1}^{m} s_i P_i, s_i \ge 0, \sum_{i=1}^{m} s_i = 1, \int \|x\|^r \, dP_i(x) < \infty$.

(a) *(Concavity)* $V_{n,r}(P) \ge \sum_{i=1}^{m} s_i V_{n,r}(P_i)$.

(b) *If* $n_i \in \mathbb{N}, \sum_{i=1}^{m} n_i \le n, \alpha_i \in C_{n_i,r}(P_i)$, *and* $\alpha = \bigcup_{i=1}^{m} \alpha_i$, *then*

$$V_{n,r}(P) \le \int \min_{a \in \alpha} \|x - a\|^r dP(x) \le \sum_{i=1}^{m} s_i V_{n_i,r}(P_i).$$

Proof

(a) Let $\alpha \in C_{n,r}(P)$. Then

$$\begin{aligned} V_{n,r}(P) &= \int \min_{a\in\alpha} \|x-a\|^r \, dP(x) \\ &= \sum_{i=1}^m s_i \int \min_{a\in\alpha} \|x-a\|^r \, dP_i(x) \\ &\geq \sum_{i=1}^m s_i V_{n,r}(P_i). \end{aligned}$$

(b) Since $|\alpha| \leq n$, we have

$$\begin{aligned} V_{n,r}(P) &\leq \int \min_{a\in\alpha} \|x-a\|^r \, dP(x) \\ &= \sum_{i=1}^m s_i \int \min_{a\in\alpha} \|x-a\|^r \, dP_i(x) \\ &\leq \sum_{i=1}^m s_i \int \min_{a\in\alpha_i} \|x-a\|^r \, dP_i(x) \\ &= \sum_{i=1}^m s_i V_{n_i,r}(P_i). \end{aligned}$$

□

Let $X = (X_1, \ldots, X_d)$.

4.15 Lemma (One-dimensional marginals)
Let the underlying norm be the l_r-norm. If $n_i \in \mathbb{N}$, $\prod_{i=1}^d n_i \leq n$, then

$$V_{n,r}(X) \leq \sum_{i=1}^d V_{n_i,r}(X_i)$$

and equality holds if and only if there exists $\alpha \in C_{n,r}(X)$ of the type $\alpha = \times_{i=1}^d \beta_i$ with $\beta_i \subset \mathbb{R}$ and $|\beta_i| = n_i$ for every i. Moreover, such n-optimal product sets α satisfy $\beta_i \in C_{n_i,r}(X_i)$ for every i.

Proof

For $1 \leq i \leq d$, let $\beta_i \subset \mathbb{R}$ with $|\beta_i| \leq n_i$ and let $\alpha = \times_{i=1}^d \beta_i$. Then $|\alpha| \leq \prod_{i=1}^d n_i \leq n$ and

$$V_{n,r}(X) \leq E \min_{a\in\alpha} \|X-a\|^r = \sum_{i=1}^d E \min_{b\in\beta_i} |X_i - b|^r.$$

Therefore, if we choose $\beta_i \in \mathcal{C}_{n_i,r}(X_i)$ for every i, we obtain

$$V_{n,r}(X) \leq \sum_{i=1}^{d} V_{n_i,r}(X_i)$$

(cf. Lemma 4.8 (b)) and if equality holds, then $\alpha \in \mathcal{C}_{n,r}(X)$. In particular, $|\alpha| = n$ by Theorem 4.1 which gives $|\beta_i| = n_i$ for every i. Conversely, assume $\alpha \in \mathcal{C}_{n,r}(X)$. Then

$$V_{n,r}(X) = \sum_{i=1}^{d} E \min_{b \in \beta_i} |X_i - b|^r \geq \sum_{i=1}^{d} V_{n_i,r}(X_i)$$

implying $V_{n,r}(X) = \sum_{i=1}^{d} V_{n_i,r}(X_i)$ and $\beta_i \in \mathcal{C}_{n_i,r}(X_i)$ for every i. □

4.3 Quantization error for ball packings

Ball packings consisting of n translates of a ball minimize the normalized n-th quantization error for bounded sets. This observation extends the corresponding statement of Lemma 2.9 for balls to the case $n \geq 2$. By a **μ-packing** in $\mathbb{R}^d$ we mean a countable family $\{C_j : j \in \mathbb{N}\}$ of Borel sets $C_j \subset \mathbb{R}^d$ such that $\mu(C_i \cap C_j) = 0$ for $i \neq j$. A λ^d-packing is simply called **packing**.

4.16 Theorem (Ball packing theorem)
Let $s > 0$ and $a_1, \ldots, a_n \in \mathbb{R}^d$ such that $\{B(a_i, s) : i = 1, \ldots, n\}$ is a packing in $\mathbb{R}^d$. Let $B = \bigcup_{i=1}^{n} B(a_i, s)$. Then

$$M_{n,r}(B) = \min\{M_{n,r}(A) : A \in \mathcal{B}(\mathbb{R}^d) \text{ bounded}, \ \lambda^d(A) > 0\}.$$

Moreover,

$$M_{n,r}(B) = n^{-r/d} M_r(B(0,1)),$$
$$\{a_1, \ldots, a_n\} \in \mathcal{C}_{n,r}(U(B)),$$

and $f = \sum_{i=1}^{n} a_i 1_{B(a_i,s)}$ is ($U(B)$-a.s. equal to) an n-optimal quantizer for $U(B)$ of order r.

Proof

Let $A \in \mathcal{B}(\mathbb{R}^d)$ be bounded with $\lambda^d(A) > 0$ and denote by Q the uniform distribution $U(A)$. Let $\mathcal{C}$ be an n-optimal partition for Q of order r. By Corollary 4.3 we know that $|\mathcal{C}| = n$ and $Q(C) > 0$ for every $C \in \mathcal{C}$. Note that $Q(\cdot|C) = U(A \cap C)$. One obtains from Lemma 2.9

$$\begin{aligned} V_{n,r}(Q) &= \sum_{C \in \mathcal{C}} V_r(Q(\cdot|C))Q(C) \\ &= \sum_{C \in \mathcal{C}} M_r(A \cap C)\lambda^d(A \cap C)^{r/d} Q(C) \\ &\geq M_r(B(0,1))\lambda^d(A)^{r/d} \sum_{C \in \mathcal{C}} Q(C)^{(d+r)/d}. \end{aligned}$$

Hölder's inequality with $p = (d+r)/d$ and $q = (d+r)/r$ gives

$$1 = \sum_{C \in \mathcal{C}} Q(C) \leq \Big(\sum_{C \in \mathcal{C}} Q(C)^p\Big)^{1/p} n^{1/q}$$

and hence

$$\sum_{C \in \mathcal{C}} Q(C)^{(d+r)/d} \geq n^{-r/d}.$$

This implies

$$V_{n,r}(Q) \geq (\lambda^d(A)/n)^{r/d} M_r(B(0,1))$$

and

$$M_{n,r}(A) \geq n^{-r/d} M_r(B(0,1)).$$

Now let $B = \bigcup_{i=1}^n B(a_i, s)$ and denote by P the uniform distribution $U(B)$. Note that $P = \frac{1}{n} \sum_{i=1}^n U(B(a_i, s))$. By Lemma 4.14 (b), we have

$$\begin{aligned} V_{n,r}(P) &\leq \int \min_{1 \leq i \leq n} \|x - a_i\|^r \, dP(x) \\ &\leq \frac{1}{n} \sum_{i=1}^n V_r(U(B(a_i, s))) \\ &= V_r(U(B(0,s))) \\ &= (\lambda^d(B)/n)^{r/d} M_r(B(0,1)). \end{aligned}$$

The last equality follows from the scale invariance of M_r (see Lemma 2.1). Thus we obtain

$$V_{n,r}(P) = \int \min_{1 \leq i \leq n} \|x - a_i\|^r \, dP(x) = V_r(U(B(0,s)))$$

and

$$M_{n,r}(B) = n^{-r/d} M_r(B(0,1)).$$

Furthermore, let $\{A_1, \ldots, A_n\}$ be a Borel measurable partition of $\mathbb{R}^d$ with $B \cap A_i \subset B(a_i, s)$ for every i and let $g = \sum_{i=1}^n a_i 1_{A_i}$. Then $f = g$ P-a.s. and since $W(a_i|\{a_1, \ldots, a_n\}) \cap B = B(a_i, s)$, $\{A_1, \ldots, A_n\}$ is a Voronoi partition with respect to $\{a_1, \ldots, a_n\}$ and P. Hence, the theorem is proved. □

4.4 Examples

We present some examples of optimal quantiziers and stationary sets of centers for dimensions $d \geq 2$. Optimal quantizers for several univariate distributions are given in the next section.

4.17 Example (Uniform distribution on a cube and the cube quantizer) Let $P = U([0,1])^d)$ and consider a tesselation of $[0,1]^d$ consisting of $n = k^d$ translates $C_1, \dots, C_n$ of the cube $[0, \frac{1}{k}]^d$. Denote by a_i the midpoint of C_i. Then $\alpha = \{a_1, \dots, a_n\} = \{\frac{2i-1}{2k} : i = 1, \dots, k\}^d$ and by the symmetry of C_i about a_i, we have $a_i \in C_r(U(C_i)) = C_r(P(\cdot|C_i))$. Let $f_n = \sum_{i=1}^{n} a_i 1_{C_i}$; see Figure 4.2. From scale and translation invariance of M_r it follows that

$$\begin{aligned} E\|X - f_n(X)\|^r &= \sum_{i=1}^{n} \int_{C_i} \|x - a_i\|^r \, dx \\ &= \sum_{i=1}^{n} M_r(C_i) P(C_i)^{(d+r)/d} \\ &= n^{-r/d} M_r([0,1]^d). \end{aligned}$$

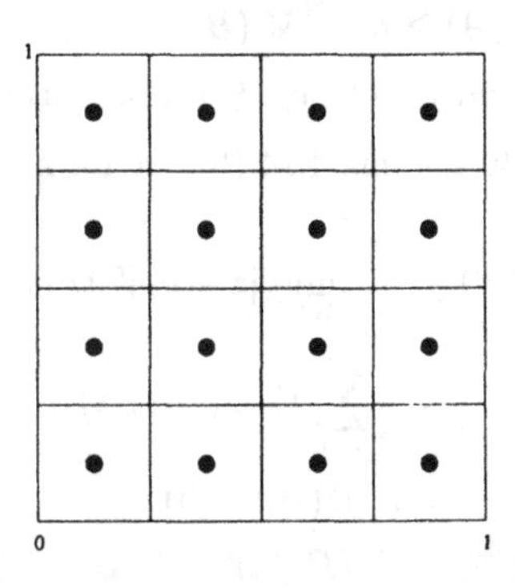

Figure 4.2: Square quantizer for $U([0,1]^2)$

We see that n-level quantization for P reduces $V_r(P) = M_r([0,1]^d)$ at least by a factor $n^{-r/d}$. Note that

$$M_r([0,1]^d) = \int_{[-\frac{1}{2},\frac{1}{2}]^d} \|x\|^r \, dx.$$

For instance, we have for the l_r-norm, $1 \leq r < \infty$

$$M_r([0,1]^d) = \frac{d}{(1+r)2^r}$$

and for the l_∞-norm

$$M_r([0,1]^d) = M_r(B(0,1)) = \frac{d}{(d+r)2^r}$$

(cf. Lemma 2.9). Note further that $\alpha \in S_{n,r}(P)$ and

$$E\|X - f_n(X)\|^r = E \min_{1 \le i \le n} \|X - a_i\|^r = \rho_r^r(P, P^{f_n})$$

provided $C_i = [0,1]^d \cap W(a_i|\alpha)$ for every i. This condition is satisfied, for instance, if the underlying norm is the l_p-norm for $1 \le p \le \infty$.

The error of the cube quantizer f_n is of optimal order $n^{-r/d}$ but the constant $M_r([0,1]^d)$ is conjectured to be not optimal for common norms with one exception. (This will be seen from the asymptotics for the n-th quantization error as $n \to \infty$ treated in Chapter II.) The exceptional case concerns the l_∞-norm in arbitrary dimensions. In this case, we have $C_i = B(a_i, \frac{1}{2k})$ and therefore, by Theorem 4.16, f_n is (P-a.s. equal to) an n-optimal quantizer of order r and $\alpha \in C_{n,r}(P)$ for every $r \ge 1$. In particular, for the l_∞-norm we obtain

$$V_{n,r}(P) = n^{-r/d} \frac{d}{(d+r)2^r}. \tag{4.3}$$

4.18 Example (Spherical distributions)
Let the underlying norm be the l_2-norm on $\mathbb{R}^d$ and let P be a spherical probability, that is, P is invariant under the orthogonal group $O(\mathbb{R}^d)$. Consider the case $r = 2$ and $n = 2$. Suppose $E\|X\|^2 < \infty$. If $\alpha = \{a_1, a_2\} \in SS_{2,2}(X)$ then $0 = EX = \sum_{i=1}^{2} a_i P(W(a_i|\alpha))$ by Remark 4.6 (b) and hence, one can find $T \in O(\mathbb{R}^d)$ such that $Ta_1 = (c_1, 0, \dots, 0)$ and $Ta_2 = (c_2, 0, \dots, 0)$ with $c_1 < 0 < c_2$. By Lemma 4.7, $T(\alpha) \in SS_{2,2}(X)$. Since

$$W(Ta_i|T(\alpha)) = W(c_i|\{c_1, c_2\}) \times \mathbb{R}^{d-1},$$

one gets $\{c_1, c_2\} \in SS_{2,2}(X_1)$. Note that X_1 is symmetric (about the origin). Now we use a uniqueness result which is discussed in the next section. Suppose the distribution of X_1 is strongly unimodal. Then it follows from Theorem 5.1 that $S_{2,2}(X_1) = \{\{-E|X_1|, E|X_1|\}\}$. Therefore, $c_1 = -E|X_1|$ and $c_2 = E|X_1|$. This yields

$$C_{2,2}(X) = SS_{2,2}(X) = \{\{-a, a\} : a \in \mathbb{R}^d, \|a\| = E|X_1|\}.$$

Since $P(W(\pm a|\{-a, a\})) = \frac{1}{2}$, it follows from Remark 4.6(c) that

$$\begin{aligned} V_{2,2}(X) &= E\|X\|^2 - (E|X_1|)^2 \\ &= dEX_1^2 - (E|X_1|)^2. \\ &= (d-1)EX_1^2 + V_{2,2}(X_1). \end{aligned}$$

4.19 Example (Uniform distribution on an euclidean ball)
Let the underlying norm be the l_2-norm on $\mathbb{R}^d$ and let $P = U(B(0,1))$. Then P is a spherical distribution. Consider the case $n = 2$ and $r = 2$. The distribution function of X_1 is given by

$$\begin{aligned} F(t) &= P([-1,t] \times \mathbb{R}^{d-1}) \\ &= \frac{1}{\lambda^d(B_d(0,1))} \int_{-1}^{t} \lambda^{d-1}\left(B_{d-1}\left(0, \sqrt{1-y^2}\right)\right) dy \\ &= \frac{\lambda^{d-1}(B_{d-1}(0,1))}{\lambda^d(B_d(0,1))} \int_{-1}^{t} (1-y^2)^{(d-1)/2} dy \\ &= \frac{\Gamma(1+\frac{d}{2})}{\sqrt{\pi}\Gamma(\frac{1}{2}+\frac{d}{2})} \int_{-1}^{t} (1-y^2)^{(d-1)/2} dy, \quad |t| \le 1. \end{aligned}$$

We thus see that X_1 has a Beta distribution. Since $\log[(1-y^2)^{(d-1)/2}]$ is concave on $(-1,1)$, the distribution of X_1 is strongly unimodal. So Example 4.18 applies. We have

$$\begin{aligned} E|X_1| &= \frac{2\Gamma(1+\frac{d}{2})}{\sqrt{\pi}\Gamma(\frac{1}{2}+\frac{d}{2})} \int_{0}^{1} y(1-y^2)^{(d-1)/2} dy \\ &= \frac{2\Gamma(1+\frac{d}{2})}{(d+1)\sqrt{\pi}\Gamma(\frac{1}{2}+\frac{d}{2})}, \\ EX_1^2 &= \frac{1}{d+2} \end{aligned}$$

and hence

$$V_{2,2}(X) = \frac{d}{d+2} - (E|X_1|)^2.$$

4.20 Example (d-dimensional standard normal distribution)
Let the underlying norm be the l_2-norm and let $P = N_d(0, I_d)$ where I_d is the unit matrix. Let $r = 2$. Here Example 4.18 applies and we obtain

$$C_{2,2}(X)) = S_{2,2}(X) = \{\{-a, a\} : a \in \mathbb{R}^d, \|a\| = \sqrt{2/\pi}\}$$

and

$$V_{2,2}(X) = d - \frac{2}{\pi}.$$

Now assume $d = 2$. For $n = 3$, we immediately find two types of 3-stationary sets of centers. Let $\alpha_1 = \gamma \times \{0\}$, where $\gamma = \{-c, 0, c\}$ with $c > 0$ is the uniquely determined 3-optimal set of centers for X_1 of order 2 (cf. Theorem 5.1). By Lemma 4.8, $\alpha_1 \in S_{3,2}(X)$ and

$$E \min_{a\in\alpha_1} \|X - a\|^2 = V_{3,2}(X_1) + 1.$$

The numerical solution is given by $c = 1.2240$ and

$$E \min_{a\in\alpha_1} \|X - a\|^2 = 1.1902$$

(cf. Table 5.1). As second configuration consider

$$\alpha_2 = \left\{(0,b), \left(\frac{\sqrt{3}b}{2}, -\frac{b}{2}\right), \left(-\frac{\sqrt{3}b}{2}, -\frac{b}{2}\right)\right\}$$

with $b > 0$. Then conv α_2 is a equilateral triangle and $P(W(a|\alpha)) = 1/3$, $a \in \alpha_2$. If b satisfies

$$\begin{aligned} b &= 3 \int_{W((0,b)|\alpha_2)} x_2 \, dN_2(0, I_2)(x) \\ &= 3 \int \int_{|x_1|/\sqrt{3}}^{\infty} x_2 \, dN(0,1)(x_2) \, dN(0,1)(x_1) \\ &= 3 \int \varphi(|x_1|/\sqrt{3}) \, dN(0,1)(x_1) \\ &= \frac{3\sqrt{3}}{2\sqrt{2\pi}} = 1.0364\ldots, \end{aligned}$$

where φ denotes the λ-density of $N(0,1)$, then $\alpha_2 \in S_{3,2}(X)$ and by Remark 4.6 (c)

$$\begin{aligned} E \min_{a\in\alpha_2} \|X - a\|^2 &= 2 - \sum_{a\in\alpha_2} \|a\|^2/3 \\ &= 2 - \frac{27}{8\pi} = 0.9257\ldots. \end{aligned}$$

Note that α_2 is considerably better than α_1. Flury (1990) provides numerical evidence that $\alpha_2 \in C_{3,2}(X)$.

For $n = 4$, we find three types of 4-stationary sets of centers. Let $\beta_1 = \gamma \times \{0\}$, where $\gamma = \{-c_2, -c_1, c_1, c_2\}$ with $0 < c_1 < c_2$ is the uniquely determined 4-optimal set of centers for X_1 of order 2 (cf. Theorem 5.1). Then by Lemma 4.8, $\beta_1 \in S_{4,2}(X)$ and

$$E \min_{a\in\beta_1} \|X - a\|^2 = V_{4,2}(X_1) + 1.$$

The numerical solution is given by $c_1 = 0.4528$, $c_2 = 1.5104$ and

$$E \min_{a\in\beta_1} \|X - a\|^2 = 1.1175$$

(cf. Table 5.1). Next consider

$$\beta_2 = \left\{ (0,0), (0,b), \left(\frac{\sqrt{3}b}{2}, -\frac{b}{2}\right), \left(-\frac{\sqrt{3}b}{2}, -\frac{b}{2}\right) \right\}$$

with $b > 0$, where b solves the equation

$$b \int_{b/2}^{\infty} (2\Phi(\sqrt{3}y) - 1)\, dN(0,1)(y) = \int_{b/2}^{\infty} (2\Phi(\sqrt{3}y) - 1) y\, dN(0,1)(y).$$

Here Φ denotes the distribution function of $N(0,1)$. Then for $a \in \beta_2, a \neq (0,0)$

$$\begin{aligned} aP(W(a|\beta_2)) &= \int_{W(a|\beta_2)} x\, dP(x), \\ P(W(a|\beta_2)) &= P(W((0,b)|\beta_2)) \\ &= \int_{b/2}^{\infty} (2\Phi(\sqrt{3}y) - 1)\, dN(0,1)(y). \end{aligned}$$

Since $\sum_{a \in \beta_2} a = (0,0)$, this implies

$$\int_{W((0,0)|\beta_2)} x dP(x) = (0,0).$$

Therefore, $\beta_2 \in S_{4,2}(X)$ and by Remark 4.6 (c)

$$\begin{aligned} E \min_{a \in \beta_2} \|X - a\|^2 &= 2 - \sum_{a \in \beta_2} \|a\|^2 P(W(a|\beta_2)) \\ &= 2 - 3b^2 \int_{b/2}^{\infty} (2\Phi(\sqrt{3}y) - 1)\, dN(0,1)(y). \end{aligned}$$

The numerical solution is given by $b = 1.2791$ and

$$E \min_{a \in \beta_2} \|X - a\|^2 = 0.8203.$$

The product quantizer $\beta_3 = \{-\sqrt{2/\pi}, \sqrt{2/\pi}\}^2$ beats β_1 and β_2. In fact, by Lemma 4.8, $\beta_3 \in S_{4,2}(X)$ and

$$E \min_{a \in \beta_3} \|X - a\|^2 = 2V_{2,2}(X_1) = 2 - \frac{4}{\pi} = 0.7267\ldots.$$

Gray and Karnin (1982) provide some numerical evidence for their conjecture that $\beta_i, i = 1,2,3$ are the only 4-stationary sets of centers of order 2 (up to l_2-isometries).

But a formal proof of this conjecture has not yet been given. Figure 4.3 shows the above stationary sets and the corresponding Voronoi tesselations. (Instead of β_3, a rotated version of β_3 is used.)

In three dimensions the product quantizer $\{-\sqrt{2/\pi}, \sqrt{2/\pi}\}^3$ can be improved upon. For $n = 8$, Gray and Karnin (1982) give three different configurations that beat the product quantizer. The authors report simulation results to show that these quantizers are superior. Iyengar and Solomon (1983) provide similar results based on numerical integration.

4.5 Stability properties and empirical versions

Stability and consistency results for the quantization problem are well known. See e.g. Pollard (1981, 1982a), Abaya and Wise (1984), Sabin and Gray (1986), Pärna (1988, 1990), Jahnke (1988), Cuesta-Albertos et al. (1988), Graf and Luschgy (1994b).

Let $\mathfrak{M}_r = \mathfrak{M}_r(\mathbb{R}^d)$ denote the set of all Borel probability measures P on $\mathbb{R}^d$ such that $\int \|x\|^r \, dP(x) < \infty$, $1 \leq r < \infty$. Recall that ρ_r is a metric on $\mathfrak{M}_r$ and for $P_k, P \in \mathfrak{M}_r$

$$\rho_r(P_k, P) \to 0$$

if and only if

$$P_k \xrightarrow{D} P \text{ (weak convergence) and } \int \|x\|^r \, dP_k(x) \to \int \|x\|^r \, dP(x)$$

(cf. Rachev and Rüschendorf, 1998, Theorem 2.6.4).

A stability property for the n-th quantization error of order r in terms of the L_r-minimal metric ρ_r follows immediately from Lemma 3.4. If $P_1, P_2 \in \mathfrak{M}_r$, then

$$|V_{n,r}(P_1)^{1/r} - V_{n,r}(P_2)^{1/r}| \leq \rho_r(P_1, P_2) \tag{4.4}$$

for every $n \in \mathbb{N}$. A stability result for n-optimal quantizing measures can also be based on ρ_r. The Hausdorff metric given by

$$d_H(A, B) = \max\{\max_{a \in A} \min_{b \in B} \|a - b\|, \max_{b \in B} \min_{a \in A} \|a - b\|\}$$

for nonempty compact subsets A, B of $\mathbb{R}^d$ is convenient for formulating a stability result for n-optimal sets of centers. Notice that

$$\begin{aligned}
&\Big|\Big(\int \min_{a \in A} \|x - a\|^r \, dP(x)\Big)^{1/r} - \Big(\int \min_{b \in B} \|x - b\|^r \, dP(x)\Big)^{1/r}\Big| \\
&\leq \Big(\int \Big|\min_{a \in A} \|x - a\| - \min_{b \in B} \|x - b\|\Big|^r \, dP(x)\Big)^{1/r} \\
&\leq \sup_{x \in \mathbb{R}^d} \Big|\min_{a \in A} \|x - a\| - \min_{b \in B} \|x - b\|\Big| \\
&\leq d_H(A, B).
\end{aligned} \tag{4.5}$$

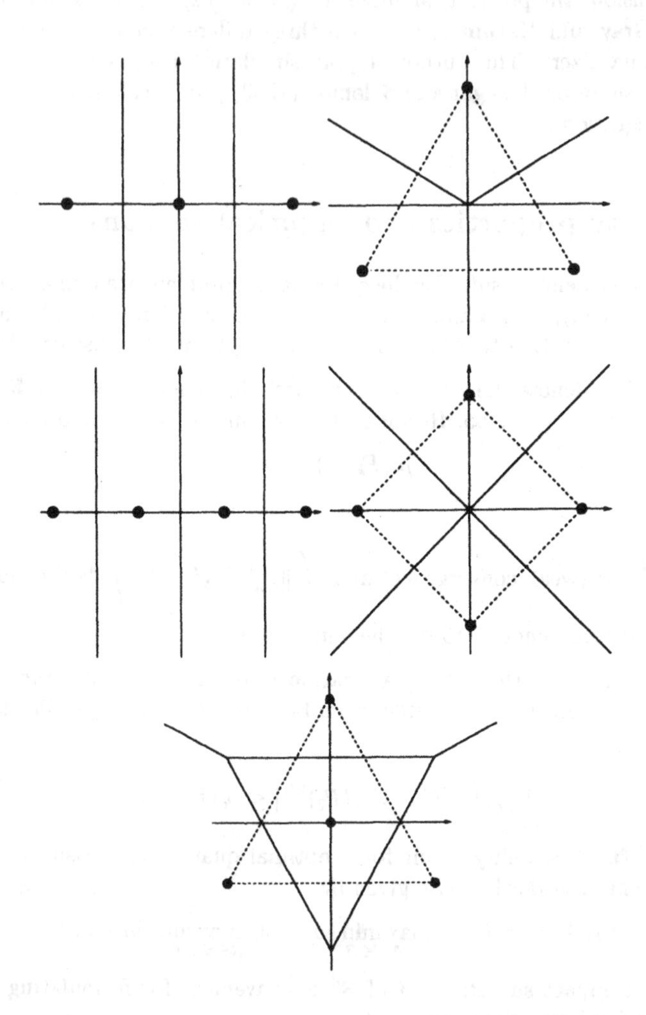

Figure 4.3: 3- and 4-stationary sets of centers for $P = N_2(0, I_2)$ of order $r = 2$ and Voronoi diagrams with respect to the l_2-norm

Let $D_{n,r}(P)$ denote the set of n-optimal quantizing measures for $P \in \mathfrak{M}_r$ of order r.

4.21 Theorem
Let $\rho_r(P_k, P) \to 0$ for P_k, $P \in \mathfrak{M}_r$ and suppose $|\text{supp}(P)| \geq n$, $n \in \mathbb{N}$.

(a) *Let $Q_k \in D_{n,r}(P_k)$, $k \in \mathbb{N}$. Then the set of ρ_r-cluster points of the sequence $(Q_k)_{k\geq 1}$ is a nonempty subset of $D_{n,r}(P)$ and*

$$\rho_r(Q_k, D_{n,r}(P)) \to 0 \text{ as } k \to \infty.$$

(b) *Let $\alpha_k \in C_{n,r}(P_k)$, $k \in \mathbb{N}$. Then the set of d_H-cluster points of the sequence $(\alpha_k)_{k\geq 1}$ is a nonempty subset of $C_{n,r}(P)$ and*

$$d_H(\alpha_k, C_{n,r}(P)) \to 0 \text{ as } k \to \infty.$$

The preceding theorem can be derived from a simple statement for arbitrary metric spaces.

4.22 Lemma
Let (M, d) be a metric space and let $N \subset M$ be a nonempty subset.

(a) *Let $f\colon N \to \mathbb{R}_+$ be a lower semicontinuous function and suppose the level set*

$$L(c) = \{y \in N\colon f(y) \leq c\}$$

is compact for some $c > \inf_{z\in N} f(z)$. Let

$$D = \{y \in N\colon f(y) = \inf_{z\in N} f(z)\}$$

and let $(y_k)_{k\geq 1}$ be a minimizing sequence in N for f, i.e., $f(y_k) \to \inf_{z\in N} f(z)$. Then the set of cluster points of $(y_k)_{k\geq 1}$ is a nonempty subset of D and

$$d(y_k, D) \to 0, \ k \to \infty.$$

(b) *For $x \in M$, set $D(x) = \{y \in N\colon d(x, y) = d(x, N)\}$. Suppose $x_k \to x$ and let $y_k \in D(x_k)$, $k \in \mathbb{N}$. Then $(y_k)_{k\geq 1}$ is a minimizing sequence in N for $f = d(x, \cdot)$.*

Proof

(a) Let $y \in M$ be the limit of a convergent subsequence $(y_{k_n})_{n\geq 1}$ of $(y_k)_{k\geq 1}$. Then $y \in L(c) \subset N$ and

$$\inf_{z\in N} f(z) = \lim f(y_{k_n}) \geq f(y).$$

We deduce $y \in D$. The existence of a cluster point of $(y_k)_{k\geq 1}$ follows immediately from the compactness of $L(c)$ and the fact that $(y_k)_{k\geq 1}$ is eventually in $L(c)$. This proves the first assertion. The second assertion is a consequence of the first one. In fact, assume $\limsup_{k\to\infty} d(y_k, D) > 0$. Then there exists $\varepsilon > 0$ and a convergent subsequence

$(y_{k_n})_{n\geq 1}$ of $(y_k)_{k\geq 1}$ satisfying $d(y_{k_n}, D) \geq \varepsilon$ for every $n \geq 1$ and $\lim y_{k_n} \in D$, a contradiction.
(b) Since

$$\begin{aligned} d(x,N) &\leq d(x,y_k) \leq d(x,x_k) + d(x_k,y_k) \\ &= d(x,x_k) + d(x_k,N) \end{aligned}$$

and $d(x_k, N) \to d(x, N)$, one gets

$$d(x, y_k) \to d(x, N) \text{ as } k \to \infty.$$

□

To prove the theorem, the following lemma is required.

4.23 Lemma
Let $B \subset \mathbb{R}^d$ be nonempty and compact. Then

$$\{\alpha \subset \mathbb{R}^d : 1 \leq |\alpha| \leq n,\ \alpha \subset B\}$$

is d_H–compact and

$$\{Q \in \mathfrak{P}_n : Q(B) = 1\}$$

is ρ_r–compact.

Proof

The d_H-compactness of $\{\alpha \subset \mathbb{R}^d : 1 \leq |\alpha| \leq n,\ \alpha \subset B\}$ follows immediately from the d_H-continuity of the map $(a_1, \ldots, a_n) \mapsto \{a_1, \ldots, a_n\}$, $a_i \in \mathbb{R}^d$. Set $\mathfrak{Q} = \{Q \in \mathfrak{P}_n : Q(B) = 1\}$. It is clear that $\mathfrak{Q}$ is relatively ρ_r-compact. To show that $\mathfrak{Q}$ is ρ_r-closed, let $(Q_k)_{k\geq 1}$ be a sequence in $\mathfrak{Q}$ and $Q \in \mathfrak{M}_r$ such that $\rho_r(Q_k, Q) \to 0$. Set $\alpha_k = \text{supp}(Q_k)$ and let $\alpha \subset \mathbb{R}^d$, $1 \leq |\alpha| \leq n$, $\alpha \subset B$ be the limit of a d_H-convergent subsequence $(\alpha_{k_j})_{j\geq 1}$ of $(\alpha_k)_{k\geq 1}$. Then for $\varepsilon > 0$ there is a $j_0 \in \mathbb{N}$ such that

$$\bigcup_{j\geq j_0} \alpha_{k_j} \subset \bigcup_{a\in\alpha} B(a,\varepsilon)$$

Since

$$\limsup_{j\to\infty} Q_{k_j}\Big(\bigcup_{a\in\alpha} B(a,\varepsilon)\Big) \leq Q\Big(\bigcup_{a\in\alpha} B(a,\varepsilon)\Big),$$

one obtains $Q(\bigcup_{a\in\alpha} B(a,\varepsilon)) = 1$. We deduce $\text{supp}(Q) \subset \alpha$ which yields $Q \in \mathfrak{Q}$. □

Proof of Theorem 4.21
(a) To show that the assertion follows from Lemma 4.22 applied to the metric space $(\mathfrak{M}_r, \rho_r)$, $N = \mathfrak{P}_n$, and $f = \rho_r(P, \cdot)$, it suffices to verify that

$$L(c) = \{Q \in \mathfrak{P}_n : \rho_r(P, Q) \leq c\}$$

is ρ_r-compact for some $c > V_{n,r}(P)^{1/r}$. Choose c such that $V_{n,r}(P)^{1/r} < c < V_{n-1,r}(P)^{1/r}$, where $V_{0,r}(P) = \infty$ (cf. Theorem 4.12). For $Q \in L(c)$ and $\alpha = \text{supp}(Q)$, we have

$$\int \min_{a\in\alpha} \|x - a\|^r \, dP(x) \le \rho_r^r(P,Q) \le c^r.$$

Hence, by Theorem 4.12 (or Lemma 2.2 in case $n = 1$)

$$L(c) \subset \{Q \in \mathfrak{P}_n : Q(B) = 1\}$$

for some compact subset B of $\mathbb{R}^d$. Using Lemma 4.23 we deduce ρ_r-compactness of $L(c)$.
(b) The assertion follows from an application of Lemma 4.22 (a) to $N = \{\alpha \subset \mathbb{R}^d : 1 \le |\alpha| \le n\}$ equipped with the Hausdorff metric d_H and $f : N \to \mathbb{R}_+$,

$$f(\alpha) = \int \min_{a\in\alpha} \|x - a\|^r \, dP(x).$$

Note first that f is d_H-continuous. This follows from (4.5). Next, consider the level set

$$L(c) = \{\alpha \in N : f(\alpha) \le c\}$$

for $V_{n,r}(P) < c < V_{n-1,r}(P)$. By Theorem 4.12 (or Lemma 2.2 in case $n = 1$), there is a compact set $B \subset \mathbb{R}^d$ such that

$$L(c) \subset \{\alpha \in N : \alpha \subset B\}.$$

Using Lemma 4.23 we deduce d_H-compactness of $L(c)$. Finally, we show that $(\alpha_k)_{k\ge1}$ is a minimizing sequence in N for f. For $k \in \mathbb{N}$, let $\{A_{k,a} : a \in \alpha_k\}$ be a Voronoi partition of $\mathbb{R}^d$ with respect to α_k. Set $Q_k = \sum_{a\in\alpha_k} P_k(A_{k,a})\delta_a$. Then $Q_k \in D_{n,r}(P_k)$ and

$$V_{n,r}(P)^{1/r} \le f(\alpha_k)^{1/r} \le \rho_r(P, Q_k).$$

Moreover, $\rho_r(P, Q_k) \to V_{n,r}(P)^{1/r}$ (cf. Lemma 4.22 (b) for $(\mathfrak{M}_r, \rho_r)$ and $N = \mathfrak{P}_n$). This implies

$$f(\alpha_k) \to V_{n,r}(P), k \to \infty.$$

Thus, we see that all assumptions of Lemma 4.22 (a) are satisfied. □

Notice that $C_{n,r}(P)$ is d_H-compact and $D_{n,r}(P)$ is ρ_r-compact provided $|\text{supp}(P)| \ge n$.

The stability results can be applied to the empirical analysis of the quantization problem. Let $X_1, X_2, \ldots$ be i.i.d. $\mathbb{R}^d$-valued random variables with distribution $P \in \mathfrak{M}_r$ and let $P_k = \frac{1}{k}\sum_{i=1}^{k} \delta_{X_i}$ be the empirical measure of $X_1, \ldots, X_k$. The empirical (sample) version of $V_{n,r}(P)$ is given by

$$V_{n,r}(P_k) = \frac{1}{k} \inf_{|\alpha|\le n} \sum_{i=1}^{k} \min_{a\in\alpha} \|X_i - a\|^r.$$

4.24 Corollary (Consistency)
Let $P \in \mathfrak{M}_r$.

(a) $V_{n,r}(P_k)^{1/r} \to V_{n,r}(P)^{1/r}$ *a.s. as $k \to \infty$ uniformly in $n \in \mathbb{N}$.*

(b) *Let $Q_k = Q_k(X_1, \ldots, X_k) \in D_{n,r}(P_k)$, $k \in \mathbb{N}$, and suppose $|\text{supp}(P)| \geq n$. Then*

$$\rho_r(Q_k, D_{n,r}(P)) \to 0 \text{ a.s., } k \to \infty.$$

(c) *Let $\alpha_k = \alpha_k(X_1, \ldots, X_k) \in C_{n,r}(P_k)$, $k \in \mathbb{N}$, and suppose $|\text{supp}(P)| \geq n$. Then*

$$d_H(\alpha_k, C_{n,r}(P)) \to 0 \text{ a.s., } k \to \infty.$$

Proof

Since $\rho_r(P_k, P) \to 0$ a.s. by the Glivenko–Cantelli theorem for ρ_r, the assertions follow from Theorem 4.21 and (4.4). □

Rates of convergence in empirical quantization can be found in Rhee and Talagrand (1989a), Linder et al. (1994), Bartlett et al. (1998) and Graf and Luschgy (1999c).

Notes

Some material on the issue of this section is contained in Gersho and Gray (1992) and Graf and Luschgy (1994a). Theorem 4.1 belongs to the folklore of this area. Theorem 4.2 seems to be new. The characterization given in Lemma 4.4 is due to Pollard (1982a) for the l_2-norm and $r = 2$. The Counterexample 4.5 is new. In case the underlying norm is the l_2-norm, the differentiability of $\psi_{n,r}$ (cf. Lemma 4.10) has been proved by Pollard (1982b) for $r = 2$ and for arbitrary r a proof is contained in Pagès (1997). Theorem 4.16 is new. Examples 4.18–4.20 on the quantization of spherical distributions and the d-dimensional standard normal distribution are essentially taken from Gray and Karnin (1982), Iyengar and Solomon (1983), Flury (1990), Tarpey et al. (1995), and Tarpey (1995). See also Tarpey (1998).

Let us mention that n-stationary sets of centers are sometimes called self-consistent sets.

The central limit problem for n-optimal empirical centers of order $r = 2$ with respect to the l_2-norm has been solved by Pollard (1982b) under a uniqueness condition for the n-optimal population centers. A central limit result in a nonregular setting has been given by Serinko and Babu (1992) for the univariate case, $d = 1$, and an extension to non-i.i.d. sampling can be found in Serinko and Babu (1995) for $d = 1$. Hartigan (1978) has conjectured the asymptotic distribution of the empirical quantization error for a special population distribution where the uniqueness condition fails but has given no proof.

Consistency results for a quantization (clustering) procedure based on a projection pursuit technique can be found in Stute and Zhu (1995). Stability and consistency results for a trimmed version of the quantization problem are contained in Cuesta-Albertos et al. (1997) and a central limit theorem for trimmed quantizers has been given by Garciá–Escudero et al. (1999).

Theorem 4.1 provides the basis for the famous Lloyd algorithm used to design quantizers. To construct an approximation to an n–stationary set of centers for P of order r the iterative method proceeds as follows:

Let $\varepsilon > 0$ be given.

Step 1.
Choose an initial set $\alpha^{(0)}$ of n points in $\mathbb{R}^d$, calculate $c_0 = E \min_{a \in \alpha^{(0)}} \|X - a\|^r$.

Step 2.
Determine a Voronoi–partition $\mathcal{A}^{(i)}$ with respect to $\alpha^{(i)}$.

Step 3.
For each set $A \in \mathcal{A}^{(i)}$ with $P(A) > 0$ choose a center a_A for the conditional probability $P(\cdot|A)$ of order r and set $\alpha^{(i+1)} = \{a_A : A \in \mathcal{A}^{(i)}\}$.

Step 4.
Calculate $e_{i+1} = E \min_{a \in \alpha^{(i+1)}} \|X - a\|^r$. If $(e_i - e_{i+1}) \leq \varepsilon e_i$ then stop. Otherwise increase i by one and repeat Step 2,3 and 4.

This algorithm was independently discovered by Steinhaus (1956) and Lloyd in 1957 (see Lloyd 1982). It is often called Lloyd's method I, since Lloyd developed a second type of algorithm (Method II) to design quantizers in the one–dimensioned case. Many people rediscovered Lloyd's method later on. For a description of the history of the algorithm we refer the reader to Gray and Neuhoff (1998). As it stands the algorithm is hard to use in practice. But if P is a discrete probability with finite support then the above algorithm can immediately be applied. The properties of Lloyd's algorithm in the context of general deterministic descent algorithms have been discussed in Sabin and Gray (1986). Recently Bouton and Pagès (1997) thoroughly investigated a constant step stochastic gradient descent algorithm for the design of quantizers which is closely related to the Kohonen algorithms used in the theory of neural networks.

Mentioning just these two algorithms for the design of quantizers is an arbitrary act since there exists a vast amount of literature concerning this subject. For a survey we refer the reader again to Gray and Neuhoff (1998).

5 Uniqueness and optimality in one dimension

In the one-dimensional case there is a reasonable criterion for the uniqueness of n-stationary sets. This immediately gives uniqueness of n-optimal sets of centers.

In this section let $d = 1$. Let X denote a real random variable with distribution P satisfying $E|X|^r < \infty$ for some $1 \le r < \infty$. The probability P is called **strongly unimodal** if $P = h\lambda$ such that $I = \{h > 0\}$ is an open (possibly unbounded) interval and $\log h$ is concave on I. Note that such distributions have all their moments finite. For this and further properties of (nondegenerate) strongly unimodal distributions we refer to Dharmadhikari and Joag-Dev (1988).

5.1 Uniqueness

The following theorem is due to Kieffer (1983). See also Trushkin (1984).

5.1 Theorem (Uniqueness)
If P is strongly unimodal, then $|S_{n,r}(P)| = 1$ for every $n \in \mathbb{N}$, $1 \le r < \infty$.

Strongly unimodal distributions are unimodal about some mode $a \in \mathbb{R}$, i.e., the λ-density h of P is increasing on $(-\infty, a)$ and decreasing on (a, ∞). Example 4.11 (as well as the subsequent Example 5.2) shows that the assertion of Theorem 5.1 may fail for unimodal distributions.

In view of Lemma 4.7, the unique n-optimal set of centers of order r for a symmetric, strongly unimodal distribution is symmetric. It is a surprising fact that symmetric, 2-stationary sets of centers may fail to be 2-optimal for symmetric, unimodal (absolutely continuous) distributions. This is illustrated by the following example taken from Abaya and Wise (1981). The same phenomenon occurs for truncated Cauchy distributions, hyper-exponential distributions and for certain variance mixtures of normal distributions. See Karlin (1982), Tarpey (1994) and Flury (1990).

5.2 Example
Let $P = h\lambda$ with

$$h(x) = \begin{cases} -\frac{|x|}{3} + \frac{5}{12}, & |x| < 1, \\ \frac{7-|x|}{72}, & 1 \le |x| < 7, \\ 0, & |x| \ge 7. \end{cases}$$

P is symmetric and unimodal about 0. Let $n = 2$ and $r = 2$. Then $V_2(P) = \operatorname{Var} X = \frac{101}{18} = 5.611\ldots$ and it is easily verified that $S_{2,2}(P) = \{\alpha_1, \alpha_2, \alpha_3\}$ with

$$\alpha_1 = \{-1, 3\},\ \alpha_2 = \{-3, 1\},\ \alpha_3 = \left\{-\frac{61}{36}, \frac{61}{36}\right\}.$$

One obtains

$$E \min_{a\in\alpha_1} |X - a|^2 = E \min_{a\in\alpha_2} |X - a|^2 = \frac{47}{18} = 2.611\ldots$$

$$E \min_{a \in \alpha_3} |X - a|^2 = \frac{3551}{1296} = 2.739\ldots$$

(Use the formula of Remark 4.6 (c).) Hence, $C_{2,2}(P) = \{\alpha_1, \alpha_2\}$ and $V_{2,2}(P) = \frac{47}{18}$; see Figure 5.1. We see that the symmetric, 2-stationary set α_3 is not 2-optimal. It is the sharp peak of the density which causes asymmetric optimal sets of centers (and prevents P from being strongly unimodal).

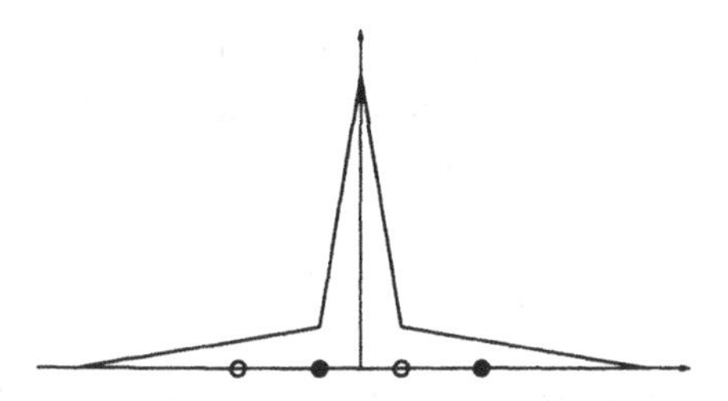

Figure 5.1: 2-optimal centers of order 2

Quantization for a symmetric distribution is related to quantization for its one-tailed version as follows.

5.3 Remark (Symmetric distributions)
Let P be symmetric with $P(\{0\}) = 0$ and $|\operatorname{supp}(P)| \geq n$. Let $Q = P(\cdot|[0, \infty))$, the one-tailed version of P.

(a) For $\alpha \subset \mathbb{R}$ and $k \in \mathbb{N}$, $\alpha \in S_{k,r}(Q)$ implies $\alpha \subset (0, \infty)$. This follows from Remark 4.6(a) and Lemma 2.5. For $\alpha \subset (0, \infty)$, one obviously obtains $\alpha \cup (-\alpha) \in S_{2k,r}(P)$ if and only if $\alpha \in S_{k,r}(Q)$. In particular, there always exists a symmetric, n-stationary set for P of order r provided n is even. Example 4.13 shows that this may fail if n is odd. (Zoppè (1997) proved the existence of symmetric, n-stationary sets for every n in case $r = 2$ under the assumption that P is absolutely continuous and $\operatorname{supp}(P)$ is convex.) Moreover, $\alpha \cup (-\alpha) \in C_{n,r}(P)$ with $n = 2k$ implies $\alpha \in C_{k,r}(Q)$ for $\alpha \subset (0, \infty)$. This follows from Theorem 4.1 since $\bigcup_{a \in \alpha} W(a|\alpha \cup (-\alpha)) = [0, \infty)$.

(b) We have $V_{n,r}(P) \leq V_{k,r}(Q)$, $n = 2k$, and equality holds if and only if there exists a symmetric set $\beta \in C_{n,r}(P)$. In this case, $\alpha \in C_{k,r}(Q)$ implies $\alpha \cup (-\alpha) \in C_{n,r}(P)$. In fact, for $\alpha \subset (0, \infty)$, $|\alpha| \leq k$,

$$\begin{aligned} V_{n,r}(P) &\leq \int \min_{b \in \alpha \cup (-\alpha)} |X - b|^r \, dP(x) \\ &= \int \min_{a \in \alpha} |X - a|^r \, dQ(x). \end{aligned}$$

Choosing $\alpha \in C_{k,r}(Q)$ gives $V_{n,r}(P) \leq V_{k,r}(Q)$ and if $V_{n,r}(P) = V_{k,r}(Q)$, then $\beta = \alpha \cup (-\alpha) \in C_{n,r}(P)$. Conversely, if $\beta \in C_{n,r}(P)$ is symmetric and $\alpha = \beta \cap (0, \infty)$, then

$$V_{n,r}(P) = \int \min_{b\in\beta} |x-b|^r \, dP(x)$$
$$= \int \min_{a\in\alpha} |x-a|^r \, dQ(x) \geq V_{k,r}(Q)$$

implying $V_{n,r}(P) = V_{k,r}(Q)$.

As yet only few examples of distributions P which are not strongly unimodal but satisfy $|S_{n,r}(P)| = 1$ are known. See Fort and Pagès (1999) and for the particular simple case $n = 2$ and $r = 2$, Yamamoto and Shinozaki (1999).

5.2 Optimal Quantizers

Let us describe optimal quantizers of orders $r = 1$ and $r = 2$ for some common univariate distributions. For $n \geq 2$, consider real numbers $a_1 < \ldots < a_n$. Let $m_i = (a_i + a_{i+1})/2$, $1 \leq i \leq n-1$, and $\alpha = \{a_1, \ldots, a_n\}$. Then the Voronoi region generated by a_i takes the form

$$W(a_1|\alpha) = (-\infty, m_1],$$
$$W(a_i|\alpha) = [m_{i-1}, m_i], \; 2 \leq i \leq n-1,$$
$$W(a_n|\alpha) = [m_{n-1}, \infty).$$

We assume that P is continuous so that the boundaries of the Voronoi regions have P-measure zero. Let F denote the distribution function of P. By Lemma 2.5, we have $\alpha \in S_{n,r}(P)$ if and only if $P(W(a_i|\alpha)) > 0$ for every i and

$$\int_{-\infty}^{a_1} (a_1 - x)^{r-1} \, dP(x) = \int_{a_1}^{m_1} (x - a_1)^{r-1} \, dP(x),$$
$$(5.1) \qquad \int_{m_{i-1}}^{a_i} (a_i - x)^{r-1} \, dP(x) = \int_{a_i}^{m_i} (x - a_i)^{r-1} \, dP(x), \; 2 \leq i \leq n-1,$$
$$\int_{m_{n-1}}^{a_n} (a_n - x)^{r-1} \, dP(x) = \int_{a_n}^{\infty} (x - a_n)^{r-1} \, dP(x).$$

In case $r = 1$, the equations (5.1) take the form

$$2F(a_1) = F(m_1),$$
$$(5.2) \qquad 2F(a_i) = F(m_i) + F(m_{i-1}), \; 2 \leq i \leq n-1,$$
$$2F(a_n) = 1 + F(m_{n-1}).$$

One obtains, for instance, $\{-F^{-1}(\frac{3}{4}), F^{-1}(\frac{3}{4})\} \in S_{2,1}(P)$ for symmetric probabilities P. ($F^{-1}(y) = \inf\{x \in I\!R : F(x) \geq y\}$, $y \in (0,1)$.)

In case $r = 2$, (5.1) takes the form

$$
\begin{aligned}
a_1 F(m_1) &= \int_{-\infty}^{m_1} x \, dP(x), \\
a_i[F(m_i) - F(m_{i-1})] &= \int_{m_{i-1}}^{m_i} x \, dP(x), 2 \leq i \leq n-1, \\
a_n[1 - F(m_{n-1})] &= \int_{m_{n-1}}^{\infty} x \, dP(x).
\end{aligned}
\tag{5.3}
$$

One obtains, for instance, $\{-E|X|, E|X|\} \in S_{2,2}(P)$ for symmetric probabilities P.

Now assume that $P = h\lambda$ is strongly unimodal. Then by Theorem 5.1, (5.1) has a unique solution in $I = \{h > 0\}$ which provides the n-optimal set of centers α for P of order r. If, additionally, P is symmetric, we have $\alpha = -\alpha$, that is, $a_i = -a_{n+1-i}$ for $1 \leq i \leq n$. Therefore, $m_k = 0$ in case $n = 2k$ and $a_{k+1} = 0$ in case $n = 2k+1$. In both cases it is enough to solve the first (or last) k equations of (5.1).

A remarkable property appears for the exponential distribution. It is the content of part (a) of the following proposition.

5.4 Proposition

(a) Let $P = E(c)$ and let $\alpha = \{a_1, \ldots, a_n\} \in C_{n,r}(P)$ with $a_1 < \ldots < a_n$. Then

$$V_{n,r}(P) = a_1^r.$$

(b) Let $P = DE(c)$ and let $\alpha = \{a_1, \ldots, a_n\} \in C_{n,r}(P)$ with $a_1 < \ldots < a_n$. Then

$$V_{n,r}(P) = a_{k+1}^r, \qquad \text{if } n = 2k,$$

$$V_{n,r}(P) = rc^r \int_0^{a_{k+2}/2c} x^{r-1} e^{-x} \, dx, \qquad \text{if } n = 2k+1.$$

Proof

We may assume without loss of generality that $c = 1$. By Theorem 5.1, α is the unique n-stationary set of order r.

(a) We have

$$\begin{aligned}
V_{n,r}(P) &= E \min_{1\le i\le n} |X-a_i|^r \\
&= \int_0^{m_1} |x-a_1|^r e^{-x}\,dx + \sum_{i=2}^{n-1} \int_{m_{i-1}}^{m_i} |x-a_i|^r e^{-x}\,dx \\
&\quad + \int_{m_{n-1}}^{\infty} |x-a_n|^r e^{-x}\,dx \\
&= \int_0^{a_1} (a_1-x)^r e^{-x}\,dx + \sum_{i=2}^{n} \int_{m_{i-1}}^{a_i} (a_i-x)^r e^{-x}\,dx \\
&\quad + \sum_{i=1}^{n-1} \int_{a_i}^{m_i} (x-a_i)^r e^{-x}\,dx + \int_{a_n}^{\infty} (x-a_n)^r e^{-x}\,dx.
\end{aligned}$$

Integration by parts yields for $b < a < c$

$$\int_b^a (a-x)^r e^{-x}\,dx = (a-b)^r e^{-b} - r\int_b^a (a-x)^{r-1} e^{-x}\,dx,$$

$$\int_a^c (x-a)^r e^{-x}\,dx = -(c-a)^r e^{-c} + r\int_a^c (x-a)^{r-1} e^{-x}\,dx,$$

$$\int_a^\infty (x-a)^r e^{-x}\,dx = r\int_a^\infty (x-a)^{r-1} e^{-x}\,dx = re^{-a}\Gamma(r).$$

Therefore

$$\begin{aligned}
V_{n,r}(P) &= a_1^r + \sum_{i=2}^{n} (a_i - m_{i-1})^r e^{-m_{i-1}} - \sum_{i=1}^{n-1} (m_i - a_i)^r e^{-m_i} \\
&\quad + r\sum_{i=1}^{n-1} \int_{a_i}^{m_i} (x-a_i)^{r-1} e^{-x}\,dx + r\int_{a_n}^{\infty} (x-a_n)^{r-1} e^{-x}\,dx - \\
&\quad - r\int_0^{a_1} (a_1-x)^{r-1} e^{-x}\,dx - r\sum_{i=2}^{n} \int_{m_{i-1}}^{a_i} (a_i-x)^{r-1} e^{-x}\,dx.
\end{aligned}$$

By (5.1), this gives

$$V_{n,r}(P) = a_1^r.$$

(b) If $n = 2k$, the assertion follows from (a) and Remark 5.3. Now let $n = 2k + 1$. Then

$$\begin{aligned} V_{n,r}(P) &= \int_0^{m_{k+1}} x^r e^{-x}\, dx + \sum_{i=2}^{k} \int_{m_{k+i-1}}^{m_{k+i}} |x - a_{k+i}|^r e^{-x}\, dx \\ &\quad + \int_{m_{n-1}}^{\infty} |x - a_n|^r e^{-x}\, dx \\ &= \int_0^{m_{k+1}} x^r e^{-x}\, dx + \sum_{i=2}^{k+1} \int_{m_{k+i-1}}^{a_{k+i}} (a_{k+i} - x)^r e^{-x}\, dx \\ &\quad + \sum_{i=2}^{k} \int_{a_{k+i}}^{m_{k+i}} (x - a_{k+i})^r e^{-x}\, dx + \int_{a_n}^{\infty} (x - a_n)^r e^{-x}\, dx. \end{aligned}$$

Again, integration by parts and (5.1) give the desired formula. □

5.5 Example (Uniform distribution)
Let $P = U([0,1])$ and let $\alpha = \{\frac{2i-1}{2n} : i = 1, \dots, n\}$. By Example 4.17, $\alpha \in C_{n,r}(P)$ for every $r \geq 1$ and

$$V_{n,r}(P) = \frac{1}{n^r (1+r) 2^r}.$$

Since P is strongly unimodal, α is the unique set of n-optimal centers of order r.

5.6 Example (Double exponential distribution)
Let $P = DE(1)$, that is, the λ-density h of P is given by $h(x) = \frac{1}{2}\exp(-|x|)$. Then $F(x) = \frac{1}{2}\exp(x)$ for $x \leq 0$ and P is symmetric and strongly unimodal. Let $r = 1$ and note that $V_1(P) = E|X| = 1$. For this distribution there exists a closed-form solution of (5.2). Let $y_i = \exp(a_i/2)$. For $n = 2k$, the first k equations of (5.2) take the form

$$\begin{aligned} 2y_1 &= y_2, & \\ 2y_i &= y_{i+1} + y_{i-1}, & 2 \leq i \leq k-1, \\ 2y_k^2 &= 1 + y_{k-1} y_k & (y_0 = 0). \end{aligned}$$

The solution of this difference equation is given by $y_i = i y_1$, $1 \leq i \leq k$, with $y_1 = (k^2 + k)^{-1/2}$. One obtains

$$\begin{aligned} a_i &= 2\log\Big(\frac{i}{\sqrt{k^2+k}}\Big), & 1 \leq i \leq k, \\ a_i &= 2\log\Big(\frac{\sqrt{k^2+k}}{n+1-i}\Big), & k+1 \leq i \leq n. \end{aligned}$$

For $n = 2k + 1$, the first k equations of (5.2) take the form

$$2y_1 = y_2,$$

$$2y_i = y_{i+1} + y_{i-1},\ 2 \le i \le k-1,$$
$$2y_k = 1 + y_{k-1}.$$

This leads to $y_i = iy_1$ with $y_1 = (k+1)^{-1}$ and hence

$$a_i = 2\log(\frac{i}{k+1}),\ 1 \le i \le k,$$
$$a_i = 2\log(\frac{k+1}{n+1-i}),\ k+2 \le i \le n.$$

In both cases $\{a_1, \dots, a_n\}$ is the unique set of n-optimal centers for P of order 1. For the quantization errror we have by Proposition 5.4

$$V_{n,1}(P) = \log(1 + \frac{2}{n}), \qquad \text{if } n \text{ is even,}$$
$$V_{n,1}(P) = \frac{2}{n+1}, \qquad \text{if } n \text{ is odd;}$$

see Table 5.9.

5.7 Example (Exponential distribution)
Let $P = E(c)$, that is, $P = h\lambda$ with $h(x) = \frac{1}{c}\exp(-\frac{x}{c})1_{(0,\infty)}(x), c > 0$. P is strongly unimodal. Let

$$a_i = 2c\log(\frac{\sqrt{n^2+n}}{n+1-i}),\ 1 \le i \le n.$$

Then $\{a_1, \dots, a_n\}$ is the unique set of n-optimal centers for P of order $r = 1$. This is a consequence of Example 5.6 and Remark 5.3 (and the scaling property of $C_{n,1}(P)$ given in Lemma 3.2 (a)). Furthermore, by Proposition 5.4

$$V_{n,1}(P) = c\log(1 + \frac{1}{n}).$$

Since $V_1(P) = E|X - c\log 2| = c\log 2$, one has to choose $c = (\log 2)^{-1}$ in order to achieve the norming $V_1(P) = 1$; see Table 5.10.

The equations (5.2) and (5.3) have been solved using MATHEMATICA for various strongly unimodal distributions. The numerical solutions are given in the Tables 5.1–5.12. (In case $P = E(1/\log 2)$, $r = 1$ and $P = DE(1)$, $r = 1$, we obtained coincidence with the exact solutions given in Examples 5.6 and 5.7 up to 5 decimal places.) The behaviour of $V_{n,r}(P)$ reflects the value of the r-th quantization coefficient of P introduced in Chapter II.

Notes

In the case of smooth densities and $r = 2$, Theorem 5.1 is due to Fleischer (1964), who provided the first uniqueness result. A proof of Theorem 5.1 (for $r = 2$) based on the "mountain pass theorem" has been given by Lamberton and Pagès (1996). See Cohort (1997) for a detailed exposition. The property of the n-th quantization error for the exponential distribution described in Proposition 5.4 is new. In the non-quantization setting $n = 1$ it has been noticed by Gilat (1988). Example 5.6 is essentially contained in Williams (1967). However, Williams intends to find n-optimal centers of order $r = 2$, but he deals with the equations (5.2) which correspond to $r = 1$. Example 5.7 is new. Tables of n-optimal sets of centers of order $r = 2$ for the normal distribution $N(0,1)$, the double exponential distribution $DE(1/\sqrt{2})$, the exponential distribution $E(1)$, and the Rayleigh distribution $W(\sqrt{2}, 2)$ (cf. Tables 5.1, 5.3–5.6) can also be found in Cox (1957), Max (1960), Lloyd (1982), Fang and Wang (1994), Adams and Giesler (1978), and Pearlman and Senge (1979).

n	1	2	3	4	5	6	7	8
a_1	0	-0.7979	-1.2240	-1.5104	-1.7241	-1.8936	-2.0334	-2.1520
a_2		0.7979	0	-0.4528	-0.7646	-1.0001	-1.1882	-1.3439
a_3			1.2240	0.4528	0	-0.3177	-0.5606	-0.7560
a_4				1.5104	0.7646	0.3177	0	-0.2451
a_5					1.7241	1.0001	0.5606	0.2451
a_6						1.8936	1.1882	0.7560
a_7							2.0334	1.3439
a_8								2.1520
$V_{n,2}$	1	0.3634	0.1902	0.1175	0.0799	0.0580	0.0440	0.0345

Table 5.1: n-optimal centers and n-th quantization error for the normal distribution $N(0,1)$ of order $r = 2$

n	1	2	3	4	5	6	7	8
a_1	0	-0.7643	-1.2621	-1.6382	-1.9422	-2.1978	-2.4185	-2.6129
a_2		0.7643	0	-0.4569	-0.7947	-1.0671	-1.2971	-1.4971
a_3			1.2621	0.4569	0	-0.3270	-0.5862	-0.8033
a_4				1.6382	0.7947	0.3270	0	-0.2548
a_5					1.9422	1.0671	0.5862	0.2548
a_6						2.1978	1.2971	0.8033
a_7							2.4185	1.4971
a_8								2.6129
$V_{n,2}$	1	0.4158	0.2307	0.1472	0.1022	0.0752	0.0576	0.0456

Table 5.2: Logistic distribution $L(\frac{\sqrt{3}}{\pi})$, $r = 2$

n	1	2	3	4	5	6	7	8
a_1	0	-0.7071	-1.4142	-1.8340	-2.2537	-2.5535	-2.8533	-3.0867
a_2		0.7071	0	-0.4198	-0.8395	-1.1393	-1.4391	-1.6725
a_3			1.4142	0.4198	0	-0.2998	-0.5996	-0.8330
a_4				1.8340	0.8395	0.2998	0	-0.2334
a_5					2.2537	1.1393	0.5996	0.2334
a_6						2.5535	1.4391	0.8330
a_7							2.8533	1.6725
a_8								3.0867
$V_{n,2}$	1	0.5000	0.2642	0.1762	0.1198	0.0899	0.0681	0.0545

Table 5.3: Double exponential distribution $DE(\frac{1}{\sqrt{2}})$, $r = 2$

n	1	2	3	4	5	6	7	8
a_1	1	0.5936	0.4240	0.3301	0.2704	0.2290	0.1986	0.1753
a_2		2.5936	1.6112	1.1780	0.9305	0.7697	0.6565	0.5725
a_3			3.6112	2.3652	1.7784	1.4298	1.1972	1.0305
a_4				4.3652	2.9657	2.2777	1.8574	1.5712
a_5					4.9657	3.4650	2.7053	2.2313
a_6						5.4650	3.8926	3.0792
a_7							5.8926	4.2665
a_8								6.2665
$V_{n,2}$	1	0.3524	0.1797	0.1090	0.0731	0.0524	0.0394	0.0307

Table 5.4: Exponential distribution $E(1)$, $r = 2$

n	1	2	3	4	5	6	7	8
a_1	1.4142	0.9271	0.7108	0.5847	0.5009	0.4407	0.3950	0.3590
a_2		2.7353	1.8420	1.4269	1.1798	1.0136	0.8932	0.8014
a_3			3.5501	2.4815	1.9577	1.6363	1.4157	1.2537
a_4				4.1445	2.9772	2.3842	2.0119	1.7523
a_5					4.6135	3.3829	2.7420	2.3325
a_6						5.0012	3.7266	3.0505
a_7							5.3318	4.0248
a_8								5.6199
$V_{n,2}$	1	0.3565	0.1836	0.1120	0.0755	0.0544	0.0410	0.0321

Table 5.5: Gamma distribution $\Gamma(\frac{1}{\sqrt{2}}, 2)$, $r = 2$

n	1	2	3	4	5	6	7	8
a_1	1.9131	1.2657	0.9772	0.8079	0.6947	0.6130	0.5508	0.5016
a_2		2.9313	2.1140	1.7010	1.4421	1.2615	1.1270	1.0222
a_3			3.4848	2.6325	2.1738	1.8745	1.6599	1.4966
a_4				3.8604	3.0025	2.5237	2.2032	1.9688
a_5					4.1425	3.2882	2.8001	2.4675
a_6						4.3670	3.5197	3.0277
a_7							4.5529	3.7136
a_8								4.7111
$V_{n,2}$	1	0.3408	0.1724	0.1042	0.0698	0.0501	0.0377	0.0294

Table 5.6: Rayleigh distribution $W(\frac{2}{\sqrt{4-\pi}}, 2)$, $r = 2$

n	1	2	3	4	5	6	7	8
a_1	0	-0.8453	-1.2898	-1.5864	-1.8067	-1.9810	-2.1244	-2.2460
a_2		0.8453	0	-0.4734	-0.7974	-1.0412	-1.2353	-1.3959
a_3			1.2898	0.4734	0	-0.3303	-0.5819	-0.7838
a_4				1.5864	0.7974	0.3303	0	-0.2540
a_5					1.8067	1.0412	0.5819	0.2540
a_6						1.9810	1.2353	0.7838
a_7							2.1244	1.3959
a_8								2.2460
$V_{n,1}$	1	0.5931	0.4258	0.3331	0.2739	0.2327	0.2024	0.1791

Table 5.7: Normal distribution $N(0, \frac{\pi}{2})$, $r = 1$

n	1	2	3	4	5	6	7	8
a_1	0	-0.7925	-1.2716	-1.6218	-1.9000	-2.1314	-2.3298	-2.5037
a_2		0.7925	0	-0.4609	-0.7925	-1.0542	-1.2716	-1.4581
a_3			1.2716	0.4609	0	-0.3265	-0.5817	-0.7925
a_4				1.6218	0.7925	0.3265	0	-0.2531
a_5					1.9000	1.0542	0.5817	0.2531
a_6						2.1314	1.2716	0.7925
a_7							2.3298	1.4581
a_8								2.5037
$V_{n,1}$	1	0.6226	0.4569	0.3620	0.3001	0.2564	0.2239	0.1988

Table 5.8: Logistic distribution $L(\frac{1}{2\log 2})$, $r = 1$

n	1	2	3	4	5	6	7	8
a_1	0	-0.6931	-1.3863	-1.7918	-2.1972	-2.4849	-2.7726	-2.9957
a_2		0.6931	0	-0.4055	-0.8109	-1.0986	-1.3863	-1.6094
a_3			1.3863	0.4055	0	-0.2877	-0.5754	-0.7985
a_4				1.7918	0.8109	0.2877	0	-0.2231
a_5					2.1972	1.0986	0.5754	0.2231
a_6						2.4849	1.3863	0.7985
a_7							2.7726	1.6094
a_8								2.9957
$V_{n,1}$	1	0.6931	0.5000	0.4055	0.3333	0.2877	0.2500	0.2231

Table 5.9: Double exponential distribution $DE(1)$, $r = 1$

n	1	2	3	4	5	6	7	8
a_1	1	0.5850	0.4150	0.3219	0.2630	0.2224	0.1926	0.1699
a_2		2.5850	1.5850	1.1520	0.9069	0.7485	0.6374	0.5552
a_3			3.5850	2.3219	1.7370	1.3923	1.1635	1
a_4				4.3219	2.9069	2.2224	1.8074	1.5261
a_5					4.9069	3.3923	2.6374	2.1699
a_6						5.3923	3.8074	3
a_7							5.8074	4.1699
a_8								6.1699
$V_{n,1}$	1	0.5850	0.4150	0.3219	0.2630	0.2224	0.1926	0.1699

Table 5.10: Exponential distribution $E(\frac{1}{\log 2})$, $r = 1$

n	1	2	3	4	5	6	7	8
a_1	1.5958	1.0650	0.8299	0.6922	0.6001	0.5344	0.4826	0.4423
a_2		2.9008	1.9829	1.5572	1.3029	1.1310	1.0058	0.9099
a_3			3.6870	2.6090	2.0824	1.7586	1.5355	1.3709
a_4				4.2542	3.0882	2.4987	2.1281	1.8688
a_5					4.6988	3.4774	2.8449	2.4405
a_6						5.0646	3.8053	3.1416
a_7							5.3754	4.0887
a_8								5.6457
$V_{n,1}$	1	0.5883	0.4195	0.3265	0.2675	0.2266	0.1966	0.1737

Table 5.11: Gamma distribution $\Gamma(a, 2)$, $a = 0.9508\ldots$, $r = 1$

n	1	2	3	4	5	6	7	8
a_1	2.2501	1.5347	1.2121	1.0204	0.8906	0.7958	0.7229	0.6647
a_2		3.3040	2.4268	1.9835	1.7041	1.5079	1.3608	1.2455
a_3			3.8697	2.9631	2.4770	2.1592	1.9304	1.7555
a_4				4.2516	3.3430	2.8389	2.5014	2.2539
a_5					4.5375	3.6352	3.1233	2.7748
a_6						4.7648	3.8714	3.3567
a_7							4.9527	4.0688
a_8								5.1124
$V_{n,1}$	1	0.5778	0.4091	0.3173	0.2594	0.2194	0.1902	0.1678

Table 5.12: Rayleigh distribution $W(a,2)$, $a = 2.7027\ldots$, $r = 1$

Chapter II

Asymptotic quantization for nonsingular probability distributions

It is difficult to find n-optimal quantizers for a fixed number n of quantizing levels at least in the multivariate case. This chapter is concerned with the theory of asymptotic quantization for nonsingular probability distributions as n tends to infinity. The asymptotic behaviour of the quantization error is derived and the asymptotic performance of certain classes of quantizers is compared with asymptotically optimal quantizers. We introduce quantization coefficients which provide interesting parameters of probability distributions. They can be evaluated for univariate distributions and some of them also for multivariate distributions. Moreover, asymptotic quantization is related to a geometric covering problem.

6 Asymptotics for the quantization error

In this section we derive the exact asymptotic first order behaviour of $V_{n,r}(P)$ (up to constants) as $n \to \infty$ in case P is not singular with respect to λ^d.

Let X be a $\mathbb{R}^d$-valued random variable with distribution P, let $\| \ \|$ denote any norm on $\mathbb{R}^d$, and let $1 \leq r < \infty$. Lemma 6.1 reflects the fact that $\bigcup_{n=1}^{\infty} \mathcal{F}_n$ is a dense subset of the Banach space $L_r(P, \mathbb{R}^d)$.

6.1 Lemma
If $E\|X\|^r < \infty$, *then*

$$\lim_{n \to \infty} V_{n,r}(P) = 0.$$

Proof

Let $\{a_1, a_2, a_3, \dots\}$ be a countable dense subset of $\mathbb{R}^d$ with $a_1 = 0$. For $\varepsilon > 0$, $\{B(a_k, (\varepsilon/2)^{1/r}) : k \in \mathbb{N}\}$ is a covering of $\mathbb{R}^d$. Therefore, one can find a Borel measurable partition $\{A_k : k \in \mathbb{N}\}$ of $\mathbb{R}^d$ satisfying $A_k \subset B(a_k, (\varepsilon/2)^{1/r})$ for every k. Choose $n \in \mathbb{N}$ such that

$$\sum_{k>n} \int_{A_k} \|x\|^r \, dP(x) \leq \varepsilon/2$$

and let $f_n = \sum_{k=1}^{n} a_k 1_{A_k}$. Then $f_n \in \mathcal{F}_n$ and

$$\begin{aligned} V_{n,r}(P) &\leq E\|X - f_n(X)\|^r \\ &= \sum_{k=1}^{n} \int_{A_k} \|x - a_k\|^r \, dP(x) + \sum_{k>n} \int_{A_k} \|x\|^r \, dP(x) \\ &\leq \varepsilon. \end{aligned}$$

□

The following theorem in its present general form is stated in Bucklew and Wise (1982) for the l_2-norm. Under some additional assumptions the result is due to Zador (1963, 1982) (who is also dealing with the l_2-norm). See also Fejes Tóth (1959) for a special case.

For a Borel measurable function $h : \mathbb{R}^d \to \mathbb{R}$ and $0 < p < \infty$ let

$$\|h\|_p = \left(\int |h|^p d\lambda^d \right)^{1/p}. \tag{6.1}$$

Furthermore, let $P = P_a + P_s$ be the Lebesgue decomposition of P with respect to λ^d, where P_a denotes the absolutely continuous part and P_s the singular part of P.

6.2 Theorem (Asymptotic quantization error)
Suppose $E\|X\|^{r+\delta} < \infty$ for some $\delta > 0$. Let

$$Q_r([0,1]^d) = \inf_{n \geq 1} n^{r/d} M_{n,r}([0,1]^d). \tag{6.2}$$

Then $Q_r([0,1]^d) > 0$ and

$$\lim_{n \to \infty} n^{r/d} V_{n,r}(P) = Q_r([0,1]^d) \left\| \frac{dP_a}{d\lambda^d} \right\|_{d/(d+r)}. \tag{6.3}$$

The proof is given below. For singular distributions, (6.3) only yields $V_{n,r}(P) = o(n^{-r/d})$ provided the above moment condition holds. An investigation of the exact order of $V_{n,r}(P)$ for several classes of singular (continuous) distributions P is contained in Chapter III.

6.3 Remark

(a) The moment condition $E\|X\|^{r+\delta} < \infty$ ensures that the limit in (6.3) is finite. In fact, $h \in L_1(\lambda^d)$, $h \geq 0$, and $\int \|x\|^{r+\delta} h(x)\, dx < \infty$ for some $\delta > 0$, implies $h \in L_{d/(d+r)}(\lambda^d)$. To see this, let $s = \frac{d}{d+r}$, $t = \frac{(r+\delta)d}{d+r}$, $p = \frac{d+r}{d}$ and $q = \frac{d+r}{r}$. Then

$$\int_{B(0,1)} h(x)^s\, dx < \infty$$

and by Hölder's inequality

$$\begin{aligned}\int_{B(0,1)^c} h(x)^s\, dx &= \int_{B(0,1)^c} h(x)^s \|x\|^t \|x\|^{-t}\, dx \\ &\leq \Big(\int_{B(0,1)^c} h(x) \|x\|^{tp}\, dx \Big)^{1/p} \Big(\int_{B(0,1)^c} \|x\|^{-tq}\, dx \Big)^{1/q} < \infty\end{aligned}$$

since $tp = r + \delta$ and $tq = \frac{(r+\delta)d}{r} > d$.

(b) The converse of the above implication does not hold. Consider, for instance,

$$h(x) = \frac{1}{2^{n(1+r)} n^{2+r}} \quad \text{if } x \in [2^n, 2^{n+1}),\ n \in \mathbb{N}.$$

Then $h \in L_1(\lambda)$ and

$$\int h^{1/(1+r)}\, d\lambda = \sum_{n=1}^{\infty} \frac{1}{n^{(2+r)/(1+r)}} < \infty$$

but

$$\begin{aligned}\int |x|^{r+\delta} h(x)\, dx &= \sum_{n=1}^{\infty} \frac{1}{2^{n(1+r)} n^{2+r}} \int_{2^n}^{2^{n+1}} |x|^{r+\delta}\, dx \\ &\geq \sum_{n=1}^{\infty} \frac{2^{n(r+\delta)} 2^n}{2^{n(1+r)} n^{2+r}} \\ &= \sum_{n=1}^{\infty} \frac{2^{n\delta}}{n^{2+r}} = \infty\end{aligned}$$

for every $\delta > 0$. Note that $\int |x|^r h(x)\, dx < \infty$.

(c) Without any moment condition we still have

$$\liminf_{n\to\infty} n^{r/d} V_{n,r}(P) \geq Q_r([0,1]^d) \left\| \frac{dP_a}{d\lambda^d} \right\|_{d/(d+r)}.$$

This is contained in the subsequent proof of Theorem 6.2 (Step 5).

The following example shows that the moment condition in Theorem 6.2 cannot be dropped.

6.4 Example
Let $x_k = 3 \cdot 2^{k-1}$ and

$$\mathbb{P}(X = x_k) = \frac{c}{2^{kr} k \log^2 k}, \; k \geq 2$$

with norming constant $c = \left(\sum_{k=2}^{\infty} \frac{1}{2^{kr} k \log^2 k}\right)^{-1}$. Then

$$EX^r = \frac{3^r c}{2^r} \sum_{k=2}^{\infty} \frac{1}{k \log^2 k} < \infty,$$

$$EX^{r+\delta} = \frac{3^{r+\delta}}{2^r} \sum_{k=2}^{\infty} \frac{2^{k\delta}}{k \log^2 k} = \infty, \; \delta > 0.$$

For $\alpha \subset \mathbb{R}$ with $|\alpha| = n$, let $I = \{k \geq 2 : \alpha \cap [2^k, 2^{k+1}) = \emptyset\}$. Then for $k \in I$

$$\min_{a \in \alpha} |x_k - a|^r \geq (x_k - 2^k)^r = 2^{(k-1)r}$$

and hence

$$\begin{aligned} E \min_{a \in \alpha} |X - a|^r &= \sum_{k=2}^{\infty} \min_{a \in \alpha} |x_k - a|^r \frac{c}{2^{kr} k \log^2 k} \\ &\geq \frac{c}{2^r} \sum_{k \in I} \frac{1}{k \log^2 k} \\ &\geq \frac{c}{2^r} \sum_{k=n+2}^{\infty} \frac{1}{k \log^2 k} \\ &\geq \frac{c}{2^r} \int_{n+2}^{\infty} \frac{1}{x \log^2 x} dx \\ &= \frac{c}{2^r \log(n+2)}. \end{aligned}$$

This gives

$$\lim_{n \to \infty} n^r V_{n,r}(X) = \infty.$$

Here the order of convergence to zero of $V_{n,r}(X)$ is $(\log n)^{-1}$. In fact, for $\beta = \{x_2, \dots, x_{n+1}\}$ one obtains

$$\min_{b \in \beta} |x_k - b|^r = 0, \; k \leq n+1,$$

$$\min_{b \in \beta} |x_k - b|^r \leq x_k^r = \frac{3^r 2^{kr}}{2^r}, \; k \geq n+2$$

and hence

$$E \min_{b \in \beta} |X - b|^r \leq \frac{3^r c}{2^r} \sum_{k=n+2}^{\infty} \frac{1}{k \log^2 k} \leq \frac{3^r c}{2^r \log(n+1)}.$$

It follows that

$$\frac{c}{2^r} \leq \liminf_{n \to \infty} \log n V_{n,r}(X)$$
$$\leq \limsup_{n \to \infty} \log n V_{n,r}(X) \leq \frac{3^r c}{2^r}.$$

In case $P_a \neq 0$ and $E\|X\|^{r+\delta} < \infty$ for some $\delta > 0$, the number

$$Q_r(P) = Q_r([0,1]^d) \left\| \frac{dP_a}{d\lambda^d} \right\|_{d/(d+r)} \tag{6.4}$$

is called **r-th quantization coefficient** of the probability P on $\mathbb{R}^d$. (Notice that $Q_r(P)$ depends on the underlying norm.) By Remark 6.3(a), $0 < Q_r(P) < \infty$. We also write $Q_r(X)$ instead of $Q_r(P)$. For $A \in \mathcal{B}(\mathbb{R}^d)$ bounded with $\lambda^d(A) > 0$, define the r-th quantization coefficient of A by

$$Q_r(A) = Q_r(U(A)). \tag{6.5}$$

Quantization coefficients provide interesting parameters of a probability distribution. They are discussed in Sections 8–10.

We shall need the following lemmas for the proof of Theorem 6.2.

6.5 Lemma
Let $P = sP_1 + (1-s)P_2$, $0 \leq s \leq 1$, $\int \|x\|^r \, dP_i(x) < \infty$. Suppose

$$n^{r/d} V_{n,r}(P_1) \to c \in [0, \infty] \text{ as } n \to \infty.$$

Then

(a)

$$\liminf_{n \to \infty} n^{r/d} V_{n,r}(P) \geq sc + (1-s) \liminf_{n \to \infty} n^{r/d} V_{n,r}(P_2),$$
$$\limsup_{n \to \infty} n^{r/d} V_{n,r}(P) \leq s(1-\varepsilon)^{-r/d} c + (1-s)\varepsilon^{-r/d} \limsup_{n \to \infty} n^{r/d} V_{n,r}(P_2)$$

for every $0 < \varepsilon < 1$.

(b) If $\lim_{n \to \infty} n^{r/d} V_{n,r}(P_2) = 0$, then

$$\lim_{n \to \infty} n^{r/d} V_{n,r}(P) = sc.$$

Proof

(a) The first inequality follows immediately from Lemma 4.14 (a). For $0 < \varepsilon < 1$, let $n_1 = n_1(n, \varepsilon) = [(1-\varepsilon)n]$ and $n_2 = n_2(n,\varepsilon) = [\varepsilon n]$. ($[x]$ denotes the integer part of $x \in \mathbb{R}$.) Then by Lemma 4.14 (b)

$$\begin{aligned} n^{r/d} V_{n,r}(P) &\le s n^{r/d} V_{n_1,r}(P_1) + (1-s) n^{r/d} V_{n_2,r}(P_2) \\ &= s(n/n_1)^{r/d} n_1^{r/d} V_{n_1,r}(P_1) + (1-s)(n/n_2)^{r/d} n_2^{r/d} V_{n_2,r}(P_2) \end{aligned}$$

for every $n \ge \max\{\frac{1}{\varepsilon}, \frac{1}{1-\varepsilon}\}$. Since $\frac{n}{n_1} \to \frac{1}{1-\varepsilon}$ and $\frac{n}{n_2} \to \frac{1}{\varepsilon}$ as $n \to \infty$, this gives the second inequality.
(b) follows from (a). □

The following result is due to Pierce (1970).

6.6 Lemma
Let $d = 1$. Then

$$V_{n,r}(P) \le n^{-r}(C_1 E|X|^{r+\delta} + C_2),\ \delta > 0,\ n \ge C_3$$

for numerical constants $C_1, C_2, C_3 > 0$ depending on δ and r (but not on P and n).

Proof

First, assume $P((1,\infty)) = 1$. We use a random quantizer argument. Consider i.i.d. Pareto-distributed random variables $Y_1, \dots, Y_n$ independent of X with distribution function

$$G(y) = \begin{cases} 1 - y^{-(\delta/r)}, & y > 1 \\ 0, & y \le 1. \end{cases}$$

Set $b = \delta/r$. Then

$$V_{n,r}(P) \le E \min_{1\le i\le n} |X - Y_i|^r = \int E \min_{1\le i\le n} |x - Y_i|^r \, dP(x)$$

and for $x > 1$

$$\begin{aligned} E \min_{1\le i\le n} |x - Y_i|^r &= r \int_0^\infty \mathbb{P}\big(\min_{1\le i\le n} |x - Y_i| > t\big) t^{r-1}\, dt \\ &= r \int_0^\infty [1 - G(x+t) + G(x-t)]^n t^{r-1}\, dt \\ &= r \int_0^{x-1} [1 - (x-t)^{-b} + (x+t)^{-b}]^n t^{r-1}\, dt \\ &\quad + r \int_{x-1}^\infty (x+t)^{-bn} t^{r-1}\, dt =: J_1(x) + J_2(x). \end{aligned}$$

Since

$$(x-t)^{-b} - (x+t)^{-b} \geq 2btx^{-(1+b)},\ 0 \leq t \leq x-1$$

and thus

$$\begin{aligned}[1-(x-t)^{-b}+(x+t)^{-b}]^n &\leq [1-2btx^{-(1+b)}]^n \\ &\leq \exp(-2nbtx^{-(1+b)}),\ 0 \leq t \leq x-1\end{aligned}$$

one gets

$$\begin{aligned} J_1(x) &\leq r\int_0^\infty \exp(-2nbtx^{-(1+b)})t^{r-1}\,dt \\ &= \frac{\Gamma(r+1)x^{r(1+b)}}{n^r(2b)^r} \\ &= \frac{\Gamma(r+1)r^r x^{r+\delta}}{n^r(2\delta)^r} =: \frac{c_1 x^{r+\delta}}{n^r},\ n \geq 1.\end{aligned}$$

Furthermore,

$$\begin{aligned} J_2(x) &\leq r\int_0^\infty (1+t)^{-bn}t^{r-1}\,dt \\ &= \frac{\Gamma(r+1)\Gamma(bn-r)}{\Gamma(bn)} \\ &\leq \frac{c_2}{n^r},\ n \geq c_3\end{aligned}$$

where the constants $c_2 \geq 1$ and $c_3 > 0$ depend only on δ and r. This gives

$$V_{n,r}(P) \leq n^{-r}(c_1 E|X|^{r+\delta} + c_2),\ \delta > 0,\ n \geq c_3.$$

Since $V_{n,r}(-X) = V_{n,r}(X)$ by Lemma 3.2, the above inequality also holds in case $P((-\infty,-1)) = 1$. If $P([-1,1]) = 1$, choose $\alpha = \{-1+\frac{2i-1}{n} : i = 1,\dots,n\}$. Then

$$V_{n,r}(P) \leq E\min_{a\in\alpha}|X-a|^r \leq n^{-r},\ n \geq 1.$$

Now let P be arbitrary. We may write

$$P = s_1 P(\cdot|(-\infty,-1)) + s_2 P(\cdot|[-1,1]) + s_3 P(\cdot|(1,\infty))$$

with $s_1 = P((-\infty,-1))$, $s_2 = P([-1,1])$, and $s_3 = P((1,\infty))$. Let $n_1 = [n/3]$. From Lemma 4.14 (b) and the above inequalities we deduce

$$V_{n,r}(P) \leq n_1^r(c_1 E|X|^{r+\delta} + c_2),\ n_1 \geq c_3.$$

This implies

$$V_{n,r}(P) \leq n^{-r}(5^r c_1 E|X|^{r+\delta} + 5^r c_2),\ n \geq 5c_3.$$

□

Lemma 6.6 is easily generalized to arbitrary dimensions d.

6.7 Corollary
We have

$$V_{n,r}(P) \leq n^{-r/d}(C_1 E\|X\|^{r+\delta} + C_2),\ \delta > 0,\ n \geq C_3$$

for numerical constants $C_1, C_2, C_3 > 0$ depending only on δ, r, d and the underlying norm.

Proof

For $p \geq 1$, let $\| \ \|_p$ denote the l_p-norm on $\mathbb{R}^d$. Recall that all norms on $\mathbb{R}^d$ are equivalent. Hence

$$V_{n,r}(P) \leq c_0 V_{n,r}(P, \| \ \|_r)$$

where $V_{n,r}(P, \| \ \|_r)$ denotes the n-th quantization error for P of order r with respect to the l_r-norm and $c_0 > 0$ is a constant depending only on r and the norm $\| \ \|$. Let $n_1 = [n^{1/d}]$. By Lemma 4.15 and Lemma 6.6

$$\begin{aligned} V_{n,r}(P, \| \ \|_r) &\leq \sum_{i=1}^{d} V_{n_1,r}(X_i) \\ &\leq n_1^{-r}\left(c_1 \sum_{i=1}^{d} E|X_i|^{r+\delta} + dc_2\right),\ \delta > 0,\ n_1 \geq c_3 \\ &\leq n^{-r/d}\left(2^r c_1 \sum_{i=1}^{d} E|X_i|^{r+\delta} + 2^r dc_2\right),\ \delta > 0,\ n \geq (2c_3)^d \end{aligned}$$

where the constants $c_1, c_2, c_3 > 0$ depend only on δ and r. Furthermore,

$$\sum_{i=1}^{d} E|X_i|^{r+\delta} = E\|X\|_{r+\delta}^{r+\delta} \leq c_4 E\|X\|^{r+\delta}$$

for a constant $c_4 > 0$ depending only on r, δ and $\| \ \|$. This proves the assertion. □

We further need an elementary lemma.

6.8 Lemma
For $m \in \mathbb{N}$ and numbers $s_i > 0$, let

$$B = \left\{(v_1, \ldots, v_m) \in (0, \infty)^m : \sum_{i=1}^{m} v_i \leq 1\right\}$$

and

$$t_i = \frac{s_i^{d/(d+r)}}{\sum_{j=1}^{m} s_j^{d/(d+r)}}, \quad 1 \le i \le m.$$

Then the function

$$F : B \to \mathbb{R}_+, \quad F(v_1, \dots, v_m) = \sum_{i=1}^{m} s_i v_i^{-r/d}$$

satisfies

$$F(t_1, \dots, t_m) = \min_{(v_1,\dots,v_m)\in B} F(v_1, \dots, v_m) = \left(\sum_{i=1}^{m} s_i^{d/(d+r)}\right)^{(d+r)/d}.$$

Moreover, $(t_1, \dots, t_m)$ *is the unique minimizer of* F.

Proof

By Hölder's inequality (for exponents less than 1) with $p = d/(d+r)$ and $q = -d/r$, one obtains

$$\begin{aligned} F(v_1, \dots, v_m) &\ge \left(\sum_{i=1}^{m} s_i^p\right)^{1/p} \left(\sum_{i=1}^{m} \left(v_i^{-r/d}\right)^q\right)^{1/q} \\ &\ge \left(\sum_{i=1}^{m} s_i^p\right)^{1/p} \\ &= F(t_1, \dots, t_m) \end{aligned}$$

for $(v_1, \dots, v_m) \in B$. To get equality $F(v_1, \dots, v_m) = F(t_1, \dots, t_m)$ it is necessary (and sufficient) that $\sum_{i=1}^{m} v_i = 1$ and $s_i = c v_i^{1/p}$, $1 \le i \le m$, for some constant $c > 0$. This implies $t_i = v_i$ for every i. □

Proof of Theorem 6.2

The proof is given by a sequence of steps from the uniform case to the general case.

Step 1. Let $P = U([0,1]^d)$. Let $m, n \in \mathbb{N}$, $m < n$ and let $k = k(n,m) = \left[\left(\frac{n}{m}\right)^{1/d}\right]$. Choose a tesselation of the unit cube $[0,1]^d$ consisting of k^d translates $C_1, \dots, C_{k^d}$ of the cube $\left[0, \frac{1}{k}\right]^d$. Then $P = k^{-d} \sum_{i=1}^{k^d} U(C_i)$. By Lemma 4.14 (b) and the translation and scale invariance of $M_{m,r}$ (see Lemma 3.2 (b)),

$$\begin{aligned} V_{n,r}(P) &\leq k^{-d} \sum_{i=1}^{k^d} V_{m,r}(U(C_i)) \\ &= k^{-d} \sum_{i=1}^{k^d} M_{m,r}(C_i) k^{-r} \\ &= k^{-r} M_{m,r}([0,1]^d) \\ &= k^{-r} V_{m,r}(P) \end{aligned}$$

and hence

$$n^{r/d} V_{n,r}(P) \leq \left(\frac{k+1}{k} \right)^r m^{r/d} V_{m,r}(P).$$

This implies

$$\limsup_{n \to \infty} n^{r/d} V_{n,r}(P) \leq m^{r/d} V_{m,r}(P)$$

for every $m \in \mathbb{N}$. Therefore, $\lim_{n \to \infty} n^{r/d} V_{n,r}(P)$ exists in $[0, \infty)$ and

$$\lim_{n \to \infty} n^{r/d} V_{n,r}(P) = Q_r([0,1]^d). \tag{6.6}$$

From Theorem 4.16 it follows that $Q_r([0,1]^d) > 0$.

Step 2. Let $P = \sum_{i=1}^{m} s_i U(C_i)$, where $\{C_1, \ldots, C_m\}$ is a packing in $\mathbb{R}^d$ consisting of closed cubes whose edges are parallel to the coordinate axes and with common length of the edges $l(C_i) = l > 0$, $s_i \geq 0$, $\sum_{i=1}^{m} s_i = 1$. Set $h = dP/d\lambda^d = \sum_{i=1}^{m} s_i l^{-d} 1_{C_i}$. We may assume without loss of generality that $s_i > 0$ for every i. For $n \in \mathbb{N}$, let

$$t_i = \frac{s_i^{d/(d+r)}}{\sum_{j=1}^{m} s_j^{d/(d+r)}} \quad \text{and } n_i = n_i(n) = [t_i n],\ 1 \leq i \leq m.$$

Then by Lemma 4.14 (b) and Lemma 3.2

$$\begin{aligned} V_{n,r}(P) &\leq \sum_{i=1}^{m} s_i V_{n_i,r}(U(C_i)) \\ &= \sum_{i=1}^{m} s_i V_{n_i,r}(U([0,1]^d)) l^r,\ n \geq \max_{1 \leq i \leq m} (1/t_i). \end{aligned}$$

From Step 1 it follows that

$$\begin{aligned} n^{r/d} V_{n_i,r}(U([0,1]^d)) &= (n/n_i)^{r/d} n_i^{r/d} V_{n_i,r}(U([0,1]^d)) \\ &\to t_i^{-r/d} Q_r([0,1]^d) \text{ as } n \to \infty. \end{aligned}$$

This implies

$$(6.7)\qquad \limsup_{n\to\infty} n^{r/d}V_{n,r}(P) \le Q_r([0,1]^d)\sum_{i=1}^m s_i t_i^{-r/d} l^r = Q_r([0,1]^d)\|h\|_{d/(d+r)}.$$

To prove that $Q_r([0,1]^d)\|h\|_{d/(d+r)}$ is a lower bound for $L = \liminf_{n\to\infty} n^{r/d}V_{n,r}(P)$, let $\beta = \beta(n) \in C_{n,r}(P)$ for $n \in \mathbb{N}$, $\beta_i = \beta_i(n) = \beta \cap \operatorname{int} C_i$, and $n_i = n_i(n) = |\beta_i|$, $1 \le i \le m$. For $0 < \varepsilon < l/2$, let $C_{i,\varepsilon} \subset C_i$ be a parallel closed cube with the same midpoint as C_i and edge-length $l(C_{i,\varepsilon}) = l - 2\varepsilon$. Choose a finite set $\gamma_i = \gamma_i(\varepsilon) \subset C_{i,\varepsilon}$, $|\gamma_i| = k = k(\varepsilon)$ such that

$$\min_{a\in\gamma_i}\|x-a\| \le \inf_{y\in C_i^c}\|x-y\| \text{ for every } x \in C_{i,\varepsilon},\ 1 \le i \le m.$$

Then

$$\begin{aligned} V_{n,r}(P) &= \sum_{i=1}^m s_i \int_{C_i} \min_{b\in\beta}\|x-b\|^r\,dx\, l^{-d} \\ &\ge \sum_{i=1}^m s_i \int_{C_{i,\varepsilon}} \min_{b\in\beta\cup\gamma_i}\|x-b\|^r\,dx\, l^{-d} \\ &= \sum_{i=1}^m s_i \int_{C_{i,\varepsilon}} \min_{b\in\beta_i\cup\gamma_i}\|x-b\|^r\,dx\, l^{-d} \\ &\ge \sum_{i=1}^m s_i V_{n_i+k,r}(U(C_{i,\varepsilon}))(l-2\varepsilon)^d l^{-d} \\ &= \sum_{i=1}^m s_i V_{n_i+k,r}(U([0,1]^d))(l-2\varepsilon)^{d+r} l^{-d},\quad n \in \mathbb{N}. \end{aligned}$$

Choose a subsequence (also denoted by (n)) such that

$$\frac{n_i}{n} \to v_i \in [0,1],\ 1 \le i \le m \text{ and } n^{r/d}V_{n,r}(P) \to L \text{ as } n \to \infty.$$

Since $\sum_{i=1}^m n_i \le n$, we have $\sum_{i=1}^m v_i \le 1$. Furthermore, $v_i > 0$ for every i. Otherwise, Step 1 yields $L = \infty$, which contradicts (6.7). By taking a further subsequence we can assume without loss of generality that

$$\lim_{n\to\infty}(n_i+k)^{r/d}V_{n_i+k,r}(U([0,1]^d)) = Q_r([0,1]^d),\ 1 \le i \le m.$$

This implies

$$L \ge Q_r([0,1]^d)\sum_{i=1}^m s_i v_i^{-r/d}(l-2\varepsilon)^{d+r}l^{-d}.$$

Since $0 < \varepsilon < l/2$ is arbitrary and by Lemma 6.8, one obtains

$$\begin{aligned} L &\geq Q_r([0,1]^d) \sum_{i=1}^{m} s_i v_i^{-r/d} l^r \\ &\geq Q_r([0,1]^d) \Big(\sum_{i=1}^{m} s_i^{d/(d+r)} \Big)^{(d+r)/d} l^r \qquad (6.8) \\ &= Q_r([0,1]^d) \|h\|_{d/(d+r)}. \end{aligned}$$

Hence, (6.3) holds in this setting.

Step 3. Let P be absolutely continuous with respect to λ^d and assume that P has a compact support. Let $\mathrm{supp}(P) \subset C$ for some closed cube C whose edges are parallel to the coordinate axes with edge-length $l(C) = l$. For $k \in \mathbb{N}$, consider a tesselation of C consisting of k^d closed cubes $C_1, \ldots, C_{k^d}$ of common edge-length $l(C_i) = l/k$. Set

$$P_k = \sum_{i=1}^{k^d} P(C_i) U(C_i),$$

$$h_k = \frac{dP_k}{d\lambda^d} = \sum_{i=1}^{k^d} \frac{P(C_i)}{\lambda^d(C_i)} 1_{C_i},$$

where $\lambda^d(C_i) = (l/k)^d$, and $h = dP/d\lambda^d$. By differentiation of measures

$$h_k \to h \quad \lambda^d\text{-a.s. as } k \to \infty$$

(cf. Cohn, 1980, Theorem 6.2.3). Therefore, by Scheffé's lemma

$$\lim_{k \to \infty} \|h_k - h\|_1 = 0.$$

Since

$$\|h_k - h\|_{d/(d+r)} \leq \lambda^d(C)^{r/d} \|h_k - h\|_1,$$

$$\left| \|h_k\|_{d/(d+r)}^{d/(d+r)} - \|h\|_{d/(d+r)}^{d/(d+r)} \right| \leq \|h_k - h\|_{d/(d+r)}^{d/(d+r)}$$

this implies

$$\lim_{k \to \infty} \|h_k\|_{d/(d+r)} = \|h\|_{d/(d+r)}.$$

Furthermore, by Step 2

$$\lim_{n \to \infty} n^{r/d} V_{n,r}(P_k) = Q_r([0,1]^d) \|h_k\|_{d/(d+r)}, \; k \in \mathbb{N}. \qquad (6.9)$$

For $n \in \mathbb{N}$ and $0 < \varepsilon < 1$, let $n_1 = n_1(n, \varepsilon) = [(1-\varepsilon)n]$ and $n_2 = n_2(n, \varepsilon) = [(\varepsilon n)^{1/d}]^d$. Consider a tesselation of C consisting of n_2 closed cubes of edge-length $l/n_2^{1/d}$ and let $\gamma = \gamma(n_2)$ denote the set of its n_2 midpoints, that is, γ corresponds to the cube n_2-quantizer for C. Then

$$\min_{a\in\gamma(n_2)} \|x-a\| \leq \frac{c_1 l}{2n_2^{1/d}} \leq \frac{c_1 l}{(\varepsilon n)^{1/d}},\ x \in C,\ n \geq 1/\varepsilon$$

for some constant $c_1 > 0$ depending only on the underlying norm. Let $\alpha(n_1, k) \in C_{n_1,r}(P_k)$ and $\delta = \delta(n,k,\varepsilon) = \alpha(n_1,k) \cup \gamma(n_2)$. Then $|\delta| \leq n$ and

$$\begin{aligned} n^{r/d}&\left|\int \min_{a\in\delta(n,k,\varepsilon)} \|x-a\|^r\, dP_k(x) - \int \min_{a\in\delta(n,k,\varepsilon)} \|x-a\|^r\, dP(x)\right| \\ &\leq n^{r/d} \int \min_{a\in\delta(n,k,\varepsilon)} \|x-a\|^r |h_k(x) - h(x)|\, dx \\ &\leq \frac{(c_1 l)^r}{\varepsilon^{r/d}} \|h_k - h\|_1 \\ &=: c_2 \|h_k - h\|_1,\ k \in \mathbb{N},\ n \geq \max\left\{\frac{1}{\varepsilon}, \frac{1}{1-\varepsilon}\right\}. \end{aligned}$$

This implies

$$\begin{aligned} n^{r/d} V_{n,r}(P) &\leq n^{r/d} \int \min_{a\in\delta(n,k,\varepsilon)} \|x-a\|^r\, dP(x) \\ &\leq n^{r/d} \int \min_{a\in\delta(n,k,\varepsilon)} \|x-a\|^r\, dP_k(x) + c_2\|h_k - h\|_1 \\ &\leq n^{r/d} \int \min_{a\in\alpha(n_1,k)} \|x-a\|^r\, dP_k(x) + c_2\|h_k - h\|_1 \\ &= n^{r/d} V_{n_1,r}(P_k) + c_2\|h_k - h\|_1,\ k \in \mathbb{N},\ n \geq \max\left\{\frac{1}{\varepsilon}, \frac{1}{1-\varepsilon}\right\} \end{aligned}$$

and hence by (6.9)

$$\limsup_{n\to\infty} n^{r/d} V_{n,r}(P) \leq (1-\varepsilon)^{-r/d} Q_r([0,1]^d) \|h_k\|_{d/(d+r)} + c_2 \|h_k - h\|_1,\ k \in \mathbb{N}.$$

Letting k tend to infinity and then letting ε tend to zero yields

$$\limsup_{n\to\infty} n^{r/d} V_{n,r}(P) \leq Q_r([0,1]^d) \|h\|_{d/(d+r)}. \tag{6.10}$$

To prove the converse estimate, let $\beta(n_1) \in C_{n_1,r}(P)$ and $\tau = \tau(n,\varepsilon) = \beta(n_1) \cup \gamma(n_2)$. Then $|\tau| \leq n$ and as above

$$\begin{aligned} n^{r/d}\left|\int \min_{a\in\tau(n,\varepsilon)} \|x-a\|^r\, dP_k(x) - \int \min_{a\in\tau(n,\varepsilon)} \|x-a\|^r\, dP(x)\right| &\leq c_2\|h_k - h\|_1, \\ k \in \mathbb{N},\ n \geq \max\left\{\frac{1}{\varepsilon}, \frac{1}{1-\varepsilon}\right\}. \end{aligned}$$

This implies

$$\begin{aligned} n^{r/d} V_{n_1,r}(P) &= n^{r/d} \int \min_{a \in \beta(n_1)} \|x-a\|^r \, dP(x) \\ &\geq n^{r/d} \int \min_{a \in \tau(n,\varepsilon)} \|x-a\|^r \, dP(x) \\ &\geq n^{r/d} \int \min_{a \in \tau(n,\varepsilon)} \|x-a\|^r \, dP_k(x) - c_2 \|h_k - h\|_1 \\ &\geq n^{r/d} V_{n,r}(P_k) - c_2\|h_k - h\|_1, \ k \in \mathbb{N}, \ n \geq \max\Big\{\frac{1}{\varepsilon}, \frac{1}{1-\varepsilon}\Big\}. \end{aligned}$$

Therefore

$$(1-\varepsilon)^{-r/d} \liminf_{n\to\infty} n_1^{r/d} V_{n_1,r}(P) \geq Q_r([0,1]^d)\|h_k\|_{d(d+r)} - c_2\|h_k - h\|_1, \ k \in \mathbb{N}.$$

Letting k tend to infinity and then letting ε tend to zero yields

$$\begin{aligned} \liminf_{n\to\infty} n^{r/d} V_{n,r}(P) &= \liminf_{n\to\infty} n_1^{r/d} V_{n_1,r}(P) \\ &\geq Q_r([0,1]^d)\|h\|_{d/(d+r)}. \end{aligned} \tag{6.11}$$

Hence, (6.3) holds in this setting.

Step 4. Let P be singular with repect to λ^d and assume that P has a compact support. Let $\text{supp}(P) \subset C$ for some open cube C whose edges are parallel to the coordinate axes with edge-length $l(C) = l$. For any $\varepsilon > 0$, there is an open set $A \subset C$ such that $P(A) = 1$ and $\lambda^d(A) \leq \varepsilon$. Moreover, there exists a countable partition $\{C_i : i \in \mathbb{N}\}$ of A consisting of half-open cubes $C_i \neq \emptyset$ with edges parallel to the coordinate axes (cf. Cohn, 1980, Lemma 1.4.2). Choose $m \in \mathbb{N}$ such that

$$l^r \sum_{i \geq m+1} P(C_i) \leq \varepsilon^{r/d}.$$

For $n \in \mathbb{N}$, let

$$s_i = \frac{\lambda^d(C_i)}{\sum_{j=1}^m \lambda^d(C_j)}, \qquad n_i = n_i(n) = [(s_i n/2)^{1/d}]^d, \qquad 1 \leq i \leq m,$$

$$n_{m+1} = n_{m+1}(n) = [(n/2)^{1/d}]^d.$$

For $1 \leq i \leq m$, consider a partition of C_i into n_i cubes of common edge-length $(\lambda^d(C_i)/n_i)^{1/d}$ and let $\alpha_i = \alpha_i(n_i)$ denote the set of its n_i midpoints. Furthermore, consider a partition of C into m_{m+1} cubes of common edge-length $l/n_{m+1}^{1/d}$ and let $\alpha_{m+1} = \alpha_{m+1}(n_{m+1})$ be the set of its n_{m+1} midpoints. Let $\alpha = \alpha(n) = \bigcup_{i=1}^{m+1} \alpha_i$. Then $|\alpha| \leq n$ and

$$\begin{aligned} \min_{a \in \alpha(n)} \|x - a\| &\leq \frac{c\lambda^d(C_i)^{1/d}}{2n_i^{1/d}} \leq \frac{c\lambda^d(C_i)^{1/d}}{(s_i n/2)^{1/d}} \\ &\leq \frac{c\varepsilon^{1/d}}{(n/2)^{1/d}}, \ x \in C_i, \ n \geq 2/s_i, \end{aligned}$$

$$\min_{a \in \alpha(n)} \|x - a\| \leq \frac{cl}{(n/2)^{1/d}}, \; x \in C, \; n \geq 2$$

for some constant $c > 0$ depending only on the underlying norm. This implies

$$\begin{aligned}
n^{r/d} V_{n,r}(P) &\leq n^{r/d} \int \min_{a \in (n)} \|x - a\|^r \, dP(x) \\
&= \sum_{i=1}^{m} n^{r/d} \int_{C_i} \min_{a \in \alpha(n)} \|x - a\|^r \, dP(x) + n^{r/d} \int_{A \setminus \cup_{i=1}^{m} C_i} \min_{a \in \alpha(n)} \|x - a\|^r \, dP(x) \\
&\leq c^r 2^{r/d} \varepsilon^{r/d} \sum_{i=1}^{m} P(C_i) + c^r 2^{r/d} l^r \sum_{i \geq m+1} P(C_i) \\
&\leq 2 c^r 2^{r/d} \varepsilon^{r/d}, n \geq \max_{1 \leq i \leq m} (2/s_i)
\end{aligned}$$

and hence

$$\limsup_{n \to \infty} n^{r/d} V_{n,r}(P) \leq 2 c^r 2^{r/d} \varepsilon^{r/d}.$$

Since $\varepsilon > 0$ is arbitrary, one obtains

$$\lim_{n \to \infty} n^{r/d} V_{n,r}(P) = 0. \tag{6.12}$$

Step 5. Let P be arbitrary. Set $h = dP_a/d\lambda^d$. For $k \in \mathbb{N}$, let $C_k = [-k, k]^d$. Then

$$P(\cdot | C_k) = \frac{h \, 1_{C_k}}{P(C_k)} \lambda^d + \frac{P_s(\cdot \cap C_k)}{P(C_k)}$$

is the Lebesgue decomposition of $P(\cdot|C_k)$ with respect to λ^d. From Steps 3 and 4 and Lemma 6.5(b) it follows

$$\lim_{n \to \infty} n^{r/d} V_{n,r}(P(\cdot|C_k)) = Q_r([0,1]^d) \left\| \frac{h \, 1_{C_k}}{P(C_k)} \right\|_{d/(d+r)}. \tag{6.13}$$

Since $V_{n,r}(P) \geq P(C_k) V_{n,r}(P(\cdot|C_k))$, this gives

$$\liminf_{n \to \infty} n^{r/d} V_{n,r}(P) \geq Q_r([0,1]^d) \|h \, 1_{C_k}\|_{d/(d+r)}.$$

Therefore, by the monotone convergence theorem as $k \to \infty$

$$\liminf_{n \to \infty} n^{r/d} V_{n,r}(P) \geq Q_r([0,1]^d) \|h\|_{d/(d+r)}. \tag{6.14}$$

Step 6. Now suppose $E\|X\|^{r+\delta} < \infty$ for some $\delta > 0$. Set $h = dP_a/d\lambda^d$. For $k \in \mathbb{N}$, let $C_k = [-k, k]^d$. Let $0 < \varepsilon < 1$. Using the decomposition $P = P(C_k)P(\cdot|C_k) + P(C_k^c)P(\cdot|C_k^c)$, it follows from (6.13) and Lemma 6.5 (a) that

$$\begin{aligned}
\limsup_{n \to \infty} n^{r/d} V_{n,r}(P) \leq{}& (1 - \varepsilon)^{-r/d} Q_r([0,1]^d) \|h \, 1_{C_k}\|_{d/(d+r)} \\
&+ P(C_k^c) \varepsilon^{-r/d} \limsup_{n \to \infty} n^{r/d} V_{n,r}(P(\cdot|C_k^c)).
\end{aligned}$$

By Corollary 6.7,

$$P(C_k^c)n^{r/d}V_{n,r}(P(\cdot|C_k^c)) \leq c_1 \int\limits_{C_k^c} \|x\|^{r+\delta}\, dP(x) + c_2 P(C_k^c),\ n \geq c_3$$

(with constants c_1, c_2, c_3 independent of k) and the above moment condition implies

$$\lim_{k\to\infty} \int\limits_{C_k^c} \|x\|^{r+\delta}\, dP(x) = 0.$$

Therefore, letting k tend to infinity in (6.15) and then letting ε tend to zero yields

$$\limsup_{n\to\infty} n^{r/d} V_{n,r}(P) \leq Q_r([0,1]^d)\|h\|_{d/(d+r)}. \tag{6.15}$$

In view of (6.14), the proof is complete. □

Notes

The approximation (6.3) of the n–th quantization error occured apparently for the first time in Panter and Dite (1951) for univariate absolutely continuous distributions and $r = 2$.

The proof of Theorem 6.2 for distributions with compact support is a simplified version (and an extension to arbitrary norms) of Bucklew and Wise (1982). The crucial point in their treatment of distributions with unbounded support is a "compander" result whose proof is not complete (cf. Linder, 1991). As shown above, the unbounded case can be resolved via the Pierce-Lemma 6.6 and its generalization to arbitrary dimensions (Corollary 6.7).

The asymptotics for empirical versions of the quantization problem (or related location problems) when both the level n and the sample size tend to infinity were studied in Hochbaum and Steele (1982), Wong (1984), Zemel (1985), Rhee and Talagrand (1989a), McGivney and Yukich (1997), Yukich (1998), Pötzelberger (1998b), and Graf and Luschgy (1999c).

7 Asymptotically optimal quantizers

Let X be a $\mathbb{R}^d$-valued random variable with distribution P, let $\| \|$ denote any norm on $\mathbb{R}^d$ and let $1 \leq r < \infty$. In the light of Theorem 6.2 a sequence $(\alpha_n)_{n\geq 1}$ with $\alpha_n \subset \mathbb{R}^d$, $|\alpha_n| \leq n$, is called **asymptotically n-optimal set of centers for P of order r** if

$$\lim_{n\to\infty} n^{r/d} E \min_{a\in\alpha_n} \|X - a\|^r = Q_r(P) \tag{7.1}$$

provided $P_a \neq 0$ and $E\|X\|^{r+\delta} < \infty$ for some $\delta > 0$. Here $Q_r(P)$ denotes the r-th quantization coefficient of P as defined in (6.4). Notice that if $(\alpha_n)_{n\geq 1}$ is asymptotically n-optimal of order r and $\{A_a : a \in \alpha_n\}$ denotes a Voronoi partition of $\mathbb{R}^d$ with respect to α_n, then $(f_n)_{n\geq 1}$ with $f_n = \sum_{a\in\alpha_n} a 1_{A_a} \in \mathcal{F}_n$ is an asymptotically n-optimal quantizer of order r, that is

$$\lim_{n\to\infty} n^{r/d} E\|X - f_n(X)\|^r = Q_r(P). \tag{7.2}$$

7.1 Mixtures and partitions

The following lemma is related to Lemma 6.8.

7.1 Lemma

Let $P = \sum_{i=1}^{m} s_i P_i$, $s_i > 0$, $\sum_{i=1}^{m} s_i = 1$, $\int \|x\|^{r+\delta} \, dP_i(x) < \infty$ for some $\delta > 0$, $P_{i,a} \neq 0$ for every i.

(a) $Q_r(P) \leq \left(\sum_{i=1}^{m} (s_i Q_r(P_i))^{d/(d+r)} \right)^{(d+r)/d}$.

(b) *Suppose*

$$Q_r(P) = \left(\sum_{i=1}^{m} (s_i Q_r(P_i))^{d/(d+r)} \right)^{(d+r)/d}. \tag{7.3}$$

For $n \in \mathbb{N}$, let

$$t_i = \frac{(s_i Q_r(P_i))^{d/(d+r)}}{\sum_{j=1}^{m} (s_j Q_r(P_j))^{d/(d+r)}}, \tag{7.4}$$

$n_i = n_i(n) = [t_i n]$, $\alpha_{i,n} \in C_{n_i,r}(P_i)$, $\alpha_n = \bigcup_{i=1}^{m} \alpha_{i,n}$. Then $(\alpha_n)_{n\geq 1}$ is an asymptotically n-optimal set of centers for P of order r.

Proof

(a) and (b). By Lemma 4.14(b)

$$\begin{aligned} V_{n,r}(P) &\le \int \min_{a \in \alpha_n} \|x-a\|^r \, dP(x) \\ &\le \sum_{i=1}^m s_i V_{n_i,r}(P_i), \; n \ge \max_{1 \le i \le m} (1/t_i) \end{aligned}$$

and by Theorem 6.2

$$\lim_{n\to\infty} n^{r/d} V_{n_i,r}(P_i) = t_i^{-r/d} Q_r(P_i), \; 1 \le i \le m.$$

This implies

$$\begin{aligned} Q_r(P) &\le \liminf_{n\to\infty} n^{r/d} \int \min_{a\in\alpha_n} \|x-a\|^r \, dP(x) \\ &\le \limsup_{n\to\infty} n^{r/d} \int \min_{a\in\alpha_n} \|x-a\|^r \, dP(x) \\ &\le \sum_{i=1}^m s_i Q_r(P_i) t_i^{-r/d} \\ &= \left(\sum_{i=1}^m (s_i Q_r(P_i))^{d/(d+r)} \right)^{(d+r)/d}. \end{aligned}$$

□

For P with $P_a = h\lambda^d \neq 0$ and $h \in L_{d/d(d+r)}(\lambda^d)$, define

(7.5) $$h_r = \frac{h^{d/(d+r)}}{\int h^{d/(d+r)} \, d\lambda^d}, \quad P_r = h_r \lambda^d.$$

Recall that by Remark 6.3, the condition $h \in L_{d/(d+r)}(\lambda^d)$ is satisfied in case $E\|X\|^{r+\delta} < \infty$ for some $\delta > 0$. The probability distribution P_r will play a central role.

7.2 Lemma
Suppose $P_a \neq 0$ and $E\|X\|^{r+\delta} < \infty$ for some $\delta > 0$. Let $\{A_1, \ldots, A_m\}$ be a P-packing in $\mathbb{R}^d$ (i.e., $P(A_i \cap A_j) = 0$ for $i \neq j$) such that $P_a(A_i) > 0$ for every i and $P(\bigcup_{i=1}^m A_i) = 1$. Then for the mixture $P = \sum_{i=1}^m P(A_i) P(\cdot|A_i)$, (7.3) is satisfied and (7.4) takes the form $t_i = P_r(A_i)$ for every i.

Proof

Let $P_a = h\lambda^d$. We have $P(\cdot|A_i)_a = h1_{A_i}\lambda^d/P(A_i) \neq 0$ and thus

$$\begin{aligned} Q_r(P(\cdot|A_i)) &= Q_r([0,1]^d)\Big(\int_{A_i} h^{d/(d+r)}\, d\lambda^d\Big)^{(d+r)/d} / P(A_i) \\ &= Q_r([0,1]^d)\Big(P_r(A_i)\int h^{d/(d+r)}\, d\lambda^d\Big)^{(d+r)/d} / P(A_i) \\ &= Q_r(P)P_r(A_i)^{(d+r)/d}/P(A_i). \end{aligned}$$

Therefore

$$(P(A_i)Q_r(P(\cdot|A_i)))^{d/(d+r)} = Q_r(P)^{d/(d+r)}P_r(A_i).$$

Since P_a and P_r are mutually absolutely continuous and hence

$$\sum_{i=1}^m P_r(A_i) = P_r\left(\bigcup_{i=1}^m A_i\right) = 1,$$

this gives the assertions. □

An interesting consequence concerns univariate symmetric and nonsingular distributions. While there are not necessarily symmetric n-optimal sets of centers (see Example 5.2 and Remark 5.3), there are symmetric asymptotically n-optimal sets of centers. More generally, we have:

7.3 Corollary (Invariant distributions)
Let G be a finite group of bijective isometries on $\mathbb{R}^d$. Suppose further P is G-invariant, $P_a \neq 0$ and $E\|X\|^{r+\delta} < \infty$ for some $\delta > 0$. Suppose there exists $A \in \mathcal{B}(\mathbb{R}^d)$ such that $\{T(A) : T \in G\}$ is a P-packing in $\mathbb{R}^d$ and $P(\bigcup_{T\in G} T(A)) = 1$. Let $n_1 = n_1(n) = [n/|G|]$ and $\alpha_n \in C_{n_1,r}(P(\cdot|A))$. Then $\big(\bigcup_{T\in G} T(\alpha_n)\big)_{n\geq 1}$ is an asymptotically n-optimal set of centers for P of order r.

Proof

For $T \in G$ we have

$$P = P^T = P_a^T + P_s^T,$$

where $P = P_a + P_s$ denotes the Lebesgue decomposition of P with respect to λ^d. Since λ^d is G-invariant, P_a^T is absolutely continuous and P_s^T is singular with respect to λ^d. So by the uniqueness of the Lebesgue decomposition, P_a is G-invariant. Let $P_a = h\lambda^d$. Then $h \circ T = h$ λ^d-a.s., $T \in G$. Therefore, P_r is also G-invariant. This implies $P_a(A) = P_a(T(A)) > 0$ and

$$P_r(A) = P_r(T(A)) = 1/|G|,\ T \in G.$$

Furthermore, since $P(\cdot|T(A)) = P(\cdot|A)^T$, $T \in G$, we obtain from Lemma 3.2. that

$$T(\alpha_n) \in C_{n_1,r}(P(\cdot|T(A))), T \in G, n \in I\!N.$$

The assertion now follows from Lemmas 7.1 and 7.2. □

7.4 Example (Sign-symmetric distributions)
Let X be sign-symmetric, that is, for every choice of signs $c_1 = \pm 1, \ldots, c_d = \pm 1$, $X = (X_1, \ldots, X_d)$ has the same distribution as $(c_1 X_1, \ldots, c_d X_d)$. The corresponding group G satisfies $|G| = 2^d$ and consists of isometries if, for instance, the underlying norm is the l_p-norm, $1 \leq p \leq \infty$. Suppose $P(\{x \in I\!R^d : x_i = 0\}) = 0$ for every i and let $A = [0, \infty)^d$. Then $\{T(A) : T \in G\}$ is a P-tesselation of $I\!R^d$.

7.2 Empirical measures

If $(\alpha_n)_{n \geq 1}$ is an asymptotically n-optimal set of centers for P of order r and $\{A_a : a \in \alpha_n\}$ denotes a Voronoi partition of $I\!R^d$ with respect to α_n, then $\left(\sum_{a \in \alpha_n} P(A_a)\delta_a\right)_{n \geq 1}$ is an asymptotically n-optimal quantizing measure of order r, that is

$$\lim_{n \to \infty} n^{1/d} \rho_r\Big(P, \sum_{a \in \alpha_n} P(A_a)\delta_a\Big) = Q_r(P)^{1/r}, \tag{7.6}$$

where ρ_r is the L_r-minimal metric defined in (3.3). In particular, $\sum_{a \in \alpha_n} P(A_a)\delta_a$ converges weakly to P.

On the other hand, α_n is asymptotically P_r-distributed in the sense that the empirical measure of α_n converges weakly to P_r. This result which is suggested by Lemma 7.2 is the content of the following theorem. A remarkable property is that P_r does not depend on the underlying norm.

7.5 Theorem
Suppose P is absolutely continuous with respect to λ^d and $E\|X\|^{r+\delta} < \infty$ for some $\delta > 0$. Let $(\alpha_n)_{n \geq 1}$ be an asymptotically n-optimal set of centers for P of order r. Then

$$\frac{1}{|\alpha_n|} \sum_{a \in \alpha_n} \delta_a \xrightarrow{D} P_r \quad \text{as } n \to \infty.$$

Proof

The proof relies on Theorem 6.2 and the equality case of Hölder's inequality (cf. Lemma 6.8). Since certainly $\lim_{n \to \infty} |\alpha_n|/n = 1$, we may assume without loss of generality that $|\alpha_n| = n$ for every $n \geq 1$. Let

$$\mu_n = \frac{1}{n} \sum_{a \in \alpha_n} \delta_a.$$

It suffices to prove that the limiting measure of any vaguely convergent subsequence of $(\mu_n)_{n\geq 1}$ coincides with P_r. Suppose for a subsequence (also denoted by (μ_n))

$$\mu_n \to \mu \text{ vaguely}$$

for some finite Borel measure μ on $\mathbb{R}^d$. Then $\mu(\mathbb{R}^d) \leq 1$. Consider a d-dimensional interval $A = (b, c]$ with $b, c \in \mathbb{R}^d$ such that $\mu(\partial A) = 0$. By vague convergence

$$\mu_n(A) \to \mu(A)$$

and hence

$$\mu_n(A^c) \to 1 - \mu(A).$$

Assume $0 < P(A) < 1$. Since P and P_r are mutually absolutely continuous, this is equivalent to $0 < P_r(A) < 1$. Write $A_1 = A$, $A_2 = A^c$, $s_i = P(A_i)$, $v_1 = \mu(A_1)$, $v_2 = 1 - \mu(A_1)$, $P_i = P(\cdot|A_i)$ and $\alpha_{i,n} = \alpha_n \cap A_i$. For $0 < \varepsilon \leq \min_{i=1,2} P_r(A_i)$, choose $b_i, c_i \in \mathbb{R}^d$, $b < b_1 < c_1 < c$, $b_2 < b < c < c_2$ such that the sets $B_1 = [b_1, c_1]$ and $B_2 = [b_2, c_2]^c$ satisfy $P(B_i) > 0$ and

$$P_r(B_i) \geq P_r(A_i) - \varepsilon, \quad i = 1, 2.$$

Then choose a finite set γ_i (on the boundary of B_i) so that

$$\min_{a\in\gamma_i} \|x - a\| \leq \inf_{y\in A_i^c} \|x - y\| \text{ for every } x \in B_i,\ i = 1, 2.$$

Say $|\gamma_i| = k$. Then we obtain

$$\begin{aligned}
\int \min_{a\in\alpha_n} \|x - a\|^r \, dP(x) &= \sum_{i=1}^{2} s_i \int \min_{a\in\alpha_n} \|x - a\|^r \, dP_i(x) \\
&\geq \sum_{i=1}^{2} s_i \int_{B_i} \min_{a\in\alpha_n\cup\gamma_i} \|x - a\|^r \, dP_i(x) \\
&= \sum_{i=1}^{2} s_i \int_{B_i} \min_{a\in\alpha_{i,n}\cup\gamma_i} \|x - a\|^r \, dP_i(x) \\
&\geq \sum_{i=1}^{2} s_i V_{n_i+k,r}(P(\cdot|B_i))P(B_i)/P(A_i),
\end{aligned}$$

where $n_i = n\mu_n(A_i) = |\alpha_{i,n}|$. This implies $v_i > 0$, $i = 1, 2$. If not, then $Q_r(P) = \infty$, a contradiction. Using Theorem 6.2 we deduce

$$\begin{aligned}
Q_r(P) &= \lim_{n\to\infty} n^{r/d} \int \min_{a\in\alpha_n} \|x - a\|^r \, dP(x) \\
&\geq \sum_{i=1}^{2} s_i v_i^{-r/d} Q_r(P(\cdot|B_i))P(B_i)/P(A_i).
\end{aligned}$$

We have

$$\begin{aligned} Q_r(P(\cdot|B_i))P(B_i) &= Q_r(P)P_r(B_i)^{(d+r)/d} \\ &\geq Q_r(P)(P_r(A_i) - \varepsilon)^{(d+r)/d}. \end{aligned}$$

Since $0 < \varepsilon \leq \min_{i=1,2} P_r(A_i)$ is arbitrary and by Lemmas 6.8 and 7.2, one obtains

$$\begin{aligned} Q_r(P) &\geq \sum_{i=1}^{2} s_i v_i^{-r/d} Q_r(P) P_r(A_i)^{(d+r)/d} / P(A_i) \\ &= \sum_{i=1}^{2} s_i Q_r(P_i) v_i^{-r/d} \\ &\geq \sum_{i=1}^{2} s_i Q_r(P_i) P_r(A_i)^{-r/d} = Q_r(P). \end{aligned}$$

Using Lemmas 6.8 and 7.2 again, this yields $v_i = P_r(A_i)$, $i = 1, 2$. Thus $\mu(A) = P_r(A)$. If $P(A) = 0$, then omit the first summands in the above considerations. One gets $v_2 = P_r(A_2) = 1$ and thus we have $\mu(A) = P_r(A) = 0$. If $P(A) = 1$, then omit the second summands. One obtains $v_1 = P_r(A_1) = 1$.

Now we have $\mu(A) = P_r(A)$ for every (bounded) d-dimensional interval $A = (b, c]$ with $\mu(\partial A) = 0$. This implies $\mu = P_r$. □

Computations of P_r can be found in Tables 7.1. and 7.2.

P	P_r
d-dimensional Normal $N_d(0, \sum)$ $\sum$ positive definite	$N_d(0, \frac{d+r}{d} \sum)$
Uniform U(B) $B \in \mathcal{B}(\mathbb{R}^d)$ bounded, $\lambda^d(B) > 0$	U(B)

Table 7.1: Probability distributions P_r

P	$(\bigotimes_1^d P)_r$
Logistic $L(a)$	$\bigotimes_1^d GL(a, \frac{d+r}{d})$
Double exponential $DE(a)$	$\bigotimes_1^d DE(\frac{a(d+r)}{d})$
Double Gamma $D\Gamma(a,b)$	$\bigotimes_1^d D\Gamma(\frac{a(d+r)}{d}, \frac{bd+r}{d+r})$
Hyper-exponential $HE(a,b)$	$\bigotimes_1^d HE(a(\frac{d+r}{d})^{1/b}, b)$
Exponential $E(a)$	$\bigotimes_1^d E(\frac{a(d+r)}{d})$
Gamma $\Gamma(a,b)$	$\bigotimes_1^d \Gamma(\frac{a(d+r)}{d}, \frac{bd+r}{d+r})$
Weibull $W(a,b)$	$\bigotimes_1^d G\Gamma(a(\frac{d+r}{d})^{1/b}, \frac{bd+r}{d+r}, b)$
Pareto $P(a,b), bd > r$	$\bigotimes_1^d P(a, \frac{bd-r}{d+r})$

Table 7.2: Probability distributions $\left(\bigotimes_1^d P\right)_r$

7.3 Asymptotic optimality in one dimension

For univariate distributions, the necessary condition of Theorem 7.5 can be turned around and used to construct asymptotically n-optimal sets of centers.

Let $d = 1$ and let $P = h\lambda$ such that $I = \{h > 0\}$ is an open (possibly unbounded) interval and h is continuous on I. Suppose $E|X|^{r+\delta} < \infty$ for some $\delta > 0$. For $n \in \mathbb{N}$, let a_i denote the $\frac{2i-1}{2n}$-quantile of P_r, $1 \leq i \leq n$, and let $m_i = (a_i + a_{i+1})/2$,

$1 \le i \le n-1$. Then $a_i \in I$ and

$$E \min_{1\le i\le n} |X - a_i|^r = \int_{-\infty}^{a_1} (a_1 - x)^r h(x)\,dx + \sum_{i=2}^{n} \int_{m_{i-1}}^{a_i} (a_i - x)^r h(x)\,dx + \sum_{i=1}^{n-1} \int_{a_i}^{m_i} (x - a_i)^r h(x)\,dx + \int_{a_n}^{\infty} (x - a_n)^r h(x)\,dx.$$

The mean value theorem yields

$$\int_{m_{i-1}}^{a_i} (a_i - x)^r h(x)\,dx = h(u_i) \int_{m_{i-1}}^{a_i} (a_i - x)^r\,dx = \frac{1}{(1+r)2^{r+1}} h(u_i)(a_i - a_{i-1})^{r+1},$$

$a_{i-1} < m_{i-1} \le u_i \le a_i$, $2 \le i \le n$,

$$\int_{a_i}^{m_i} (x - a_i)^r h(x)\,dx = h(v_{i+1}) \int_{a_i}^{m_i} (x - a_i)^r\,dx = \frac{1}{(1+r)2^{r+1}} h(v_{i+1})(a_{i+1} - a_i)^{r+1},$$

$a_i \le v_{i+1} \le m_i < a_{i+1}$, $1 \le i \le n-1$,

$$\frac{1}{n} = \frac{2i-1}{2n} - \frac{2(i-1)-1}{2n} = \int_{a_{i-1}}^{a_i} h_r(x)\,dx = h_r(w_i)(a_i - a_{i-1}),$$

$a_{i-1} \le w_i \le a_i$, $2 \le i \le n$.

This gives

$$E \min_{1\le i\le n} |X - a_i|^r = \int_{-\infty}^{a_1} (a_1 - x)^r h(x)\,dx + \int_{a_n}^{\infty} (x - a_n)^r h(x)\,dx + \frac{1}{(1+r)2^r n^r} \Big[\frac{1}{2} \sum_{i=2}^{n} \frac{h(u_i)}{h_r(w_i)^r}(a_i - a_{i-1}) + \frac{1}{2} \sum_{i=2}^{n} \frac{h(v_i)}{h_r(w_i)^r}(a_i - a_{i-1})\Big].$$

Note that a_i, u_i, v_i, w_i depend on n. Under suitable assumptions on the density h, we have

$$\lim_{n\to\infty} \sum_{i=2}^{n} \frac{h(y_i)}{h_r(w_i)^r}(a_i - a_{i-1}) = \int \frac{h}{(h_r)^r}\, d\lambda = \|h\|_{1/(1+r)},$$

$y_i \in \{u_i, v_i\}$, and the two remaining summands are of order $o(n^{-r})$. One obtains

$$\lim_{n\to\infty} n^r E \min_{1\leq i\leq n} |X - a_i|^r = \frac{1}{(1+r)2^r}\|h\|_{1/(1+r)} = Q_r(P). \tag{7.7}$$

For the latter equation see Example 5.5. Thus $(\{a_1, \ldots, a_n)\})_{n\geq 1}$ is asymptotically n-optimal for P of order r. The same result holds for the $\frac{i}{n+1}$-quantiles of P_r, $1 \leq i \leq n$.

Sufficient conditions for the above result to be valid can be found in Cambanis and Gerr (1983) (with a gap in the proof), Linder (1991), Pötzelberger and Felsenstein (1994).

7.6 Example (Double exponential distribution)
Let $P = DE(c)$. The $\frac{i}{n+1}$-quantiles of $P_r = DE((1+r)c)$ are given by

$$a_i = (1+r)c\log(\frac{2i}{n+1}),\ 1 \leq i \leq \frac{n+1}{2},$$
$$a_i = (1+r)c\log(\frac{n+1}{2n+2-2i}),\ \frac{n+1}{2} \leq i \leq n.$$

In case n odd and $r = 1$, they coincide with the n-optimal set of centers for P of order 1 (cf. Example 5.6).

The error of quantizers of the above type for various distributions is evaluated in Tables 7.3 and 7.4. These values should be compared with the n-th quantization error given in Tables 5.1 - 5.12.

$P\setminus n$	2	3	4	5	6	7	8
$N(0,1)$	0.5006	0.2466	0.1456	0.0960	0.0680	0.0506	0.0392
	0.3661	0.1913	0.1180	0.0802	0.0581	0.0441	0.0346
$L(\frac{\sqrt{3}}{\pi})$	0.7332	0.3373	0.1978	0.1304	0.0925	0.0690	0.0535
	0.4188	0.2319	0.1478	0.1026	0.0754	0.0578	0.0457
$DE(\frac{1}{\sqrt{2}})$	1.0826	0.3657	0.2418	0.1508	0.1112	0.0811	0.0641
	0.5234	0.2648	0.1782	0.1200	0.0903	0.0682	0.0546
$E(1)$	0.4836	0.2223	0.1282	0.0834	0.0586	0.0434	0.0335
	0.6112	0.2763	0.1551	0.0987	0.0680	0.0497	0.0378
$\Gamma(\frac{1}{\sqrt{2}},2)$	0.4625	0.2232	0.1311	0.0861	0.0609	0.0453	0.0350
	0.4761	0.2269	0.1323	0.0866	0.0610	0.0453	0.0350
$W(\frac{2}{\sqrt{4-\pi}},2)$	0.4214	0.2040	0.1196	0.0784	0.0554	0.0411	0.0318
	0.3896	0.1935	0.1151	0.0762	0.0541	0.0404	0.0313

Table 7.3: $r = 2$, $V_2(P) = 1$. Quantization error for $\frac{2i-1}{2n}$-quantiles (first line) and $\frac{i}{n+1}$-quantiles (second line) of P_2, $1 \leq i \leq n$

$P\setminus n$	2	3	4	5	6	7	8
$N(0,\frac{\pi}{2})$	0.6512	0.4613	0.3565	0.2904	0.2450	0.2119	0.1866
	0.5965	0.4279	0.3345	0.2748	0.2334	0.2029	0.1795
$L(\frac{1}{2\log 2})$	0.7284	0.5151	0.3987	0.3253	0.2749	0.2380	0.2099
	0.6226	0.4569	0.3620	0.3001	0.2564	0.2239	0.1988
$DE(1)$	0.8863	0.5556	0.4504	0.3600	0.3091	0.2653	0.2358
	0.6998	0.5000	0.4063	0.3333	0.2879	0.2500	0.2232
$E(\frac{1}{\log 2})$	0.6497	0.4459	0.3402	0.2752	0.2310	0.1991	0.1749
	0.6890	0.4694	0.3553	0.2856	0.2387	0.2050	0.1796
$\Gamma(a,2)$	0.6396	0.4476	0.3443	0.2797	0.2356	0.2035	0.1791
$a = 0.9508...$	0.6351	0.4434	0.3409	0.2771	0.2335	0.2017	0.1776
$W(a,2)$	0.6177	0.4324	0.3323	0.2698	0.2270	0.1960	0.1724
$a = 2.7027...$	0.5990	0.4222	0.3260	0.2655	0.2240	0.1937	0.1706

Table 7.4: $r = 1$, $V_1(P) = 1$. Quantization error for $\frac{2i-1}{2n}$-quantiles (first line) and $\frac{i}{n+1}$-quantiles (second line) of P_1, $1 \leq i \leq n$

7.4 Product quantizers

We will compare the asymptotic performance of n-optimal product quantizers with n-optimal quantizers. The comparison is based on the relation between $Q_r(X)$ and $Q_r(X_i)$, $1 \leq i \leq d$. For this, the following lemma is useful.

7.7 Lemma
Let

$$B = \left\{ (v_1, \ldots, v_d) \in (0, \infty)^d : \prod_{i=1}^{d} v_i \leq 1 \right\}$$

and for $s_i > 0$ *let*

$$t_i = \frac{s_i^{1/r}}{\prod_{j=1}^{d} s_j^{1/rd}}, \quad 1 \leq i \leq d.$$

Then the function

$$F : B \to \mathbb{R}_+, \quad F(v_1, \ldots, v_d) = \sum_{i=1}^{d} s_i v_i^{-r}$$

satisfies

$$F(t_1, \ldots, t_d) = \min_{(v_1, \ldots, v_d) \in B} F(v_1, \ldots, v_d) = d \prod_{i=1}^{d} s_i^{1/d}.$$

Proof

By the arithmetic-geometric mean inequality one obtains

$$\begin{aligned} F(v_1, \ldots, v_d) &\geq d \Big(\prod_{i=1}^{d} s_i v_i^{-r} \Big)^{1/d} \\ &\geq d \prod_{i=1}^{d} s_i^{1/d} \\ &= F(t_1, \ldots, t_d) \end{aligned}$$

for $(v_1, \ldots, v_d) \in B$. □

7.8 Lemma
Let the underlying norm be the l_r*-norm with* $1 \leq r < \infty$. *Suppose* $E\|X\|^{r+\delta} < \infty$ *for some* $\delta > 0$ *and* $Q_r(X_i) > 0$ *for* $1 \leq i \leq d$.

(a) $Q_r(X) \leq d \prod_{i=1}^{d} Q_r(X_i)^{1/d}$.

(b) *If* $t_i = Q_r(X_i)^{1/r} / \prod_{j=1}^{d} Q_r(X_j)^{1/rd}$, $n_i = [t_i n^{1/d}]$ *for* $n \in \mathbb{N}$, $\beta_{i,n} \in C_{n_i,r}(X_i)$ *and* $\alpha_n = \times_{i=1}^{d} \beta_{i,n}$, *then*

$$\lim_{n \to \infty} n^{r/d} E \min_{a \in \alpha_n} \|X - a\|^r = d \prod_{i=1}^{d} Q_r(X_i)^{1/d}.$$

Proof

The choice of t_i comes from Lemma 7.7. We have

$$\begin{aligned} V_{n,r}(X) &\leq E \min_{a \in \alpha_n} \|X - a\|^r \\ &= \sum_{i=1}^d E \min_{b \in \beta_{i,n}} |X_i - b|^r = \sum_{i=1}^d V_{n_i,r}(X_i) \end{aligned}$$

provided $n_i \geq 1$ for every i. Since by Theorem 6.2

$$n^{r/d} V_{n_i,r}(X_i) = \left(\frac{n^{1/d}}{n_i}\right)^r n_i^r V_{n_i,r}(X_i) \to t_i^{-r} Q_r(X_i)$$

as $n \to \infty$, one obtains

$$\begin{aligned} Q_r(X) &\leq \lim_{n \to \infty} n^{r/d} E \min_{a \in \alpha_n} \|X - a\|^r \\ &= \sum_{i=1}^n t_i^{-r} Q_r(X_i) \\ &= d \prod_{i=1}^d Q_r(X_i)^{1/d}. \end{aligned}$$

□

In view of Lemmas 7.7 and 7.8, the number

$$\frac{d \prod_{i=1}^d Q_r(X_i)^{1/d}}{Q_r(X_1, \ldots, X_d)} \tag{7.8}$$

represents the "vector quantizer advantage" provided the underlying norm is the l_r-norm and assumptions of Lemma 7.8 are satisfied. For the d-asymptotics of the vector quantizer advantage see Remark 9.6 (a).

7.9 Remark

Suppose $E\|X\|^{r+\delta} < \infty$ for some $\delta > 0$. Suppose further that the one-dimensional marginal distributions P_i of P are absolutely continuous with respect to λ, $1 \leq i \leq d$. Let $n_i = n_i(n) \in I\!N$ such that $\prod_{i=1}^d n_i \leq n$, let $(\beta_{i,n})_{n \geq 1}$ be an asymptotically n_i-optimal set of centers for P_i of order r and let $\alpha_n = \times_{i=1}^d \beta_{i,n}$. Then as $n \to \infty$

$$\frac{1}{|\alpha_n|} \sum_{a \in \alpha_n} \delta_a \xrightarrow{D} \bigotimes_{i=1}^d P_{i,r}.$$

In fact, we have

$$\frac{1}{|\alpha_n|} \sum_{a \in \alpha_n} \delta_a = \bigotimes_{i=1}^d \left(\frac{1}{|\beta_{i,n}|} \sum_{b \in \beta_{i,n}} \delta_b \right)$$

and by Theorem 7.5

$$\frac{1}{|\beta_{i,n}|} \sum_{b \in \beta_{i,n}} \delta_b \xrightarrow{D} P_{i,r}, \ 1 \leq i \leq d.$$

Note that in case $d \geq 2$

$$\left(\bigotimes_{i=1}^{d} P_i \right)_r \neq \bigotimes_{i=1}^{d} P_{i,r}$$

unless P_i is an uniform distribution for every i.

Notes

The observation in Corollary 7.3 seems to be new. Theorem 7.5 extends considerably a corresponding result of McClure (1975) for one-dimensional distributions with compact support. See also the review article by McClure (1980). Rates of convergence in Theorem 7.5 for some one–dimensional distributions P (Pareto, exponential and power–function distributions) with respect to various local distances have been computed by Fort and Pagès (1999). A discussion of the vector quantizer advantage as defined in (7.8) can be found in Lookabaugh and Gray (1989). Na and Neuhoff (1995) provide a nice result about the asymptotic performance of suboptimal quantizers (like product quantizers).

8 Regular quantizers and quantization coefficients

As before, let X be a $\mathbb{R}^d$-valued random variable with distribution P, let $\|\ \|$ denote any norm on $\mathbb{R}^d$, and let $1 \leq r < \infty$. The r-th quantization coefficient of P as defined in (6.4) consists of a geometric part $Q_r([0,1])^d)$ which depends only on r and the dimension d (and on the underlying norm) and a second part $\|\frac{dP_a}{d\lambda^d}\|_{d/(d+r)}$ which is related to the distribution P. For instance, the r-th quantization coefficient of a d-dimensional normal distribution $N_d(0,\Sigma)$ with positive definite covariance matrix Σ is given by

$$Q_r(N_d(0,\Sigma)) = Q_r([0,1]^d)(2\pi)^{r/2}(\frac{d+r}{d})^{(d+r)/2}(\det \Sigma)^{r/2d}.$$

The constants $Q_r([0,1]^d)$ are only known for $d=1$, $d=2$ (l_1-norm, l_2-norm), and in "trivial" cases for $d \geq 3$. It appears to be a rather challenging problem to determine the values of these constants.

First, note that the scaling property of $V_{n,r}(P)$ carries over to $Q_r(P)$.

8.1 Lemma
Let $T\colon \mathbb{R}^d \to \mathbb{R}^d$ be a similarity transformation with scaling number $c > 0$.

(a) If $P_a \neq 0$ and $E\|X\|^{r+\delta} < \infty$ for some $\delta > 0$, then

$$Q_r(T(X)) = c^r Q_r(X).$$

(b) If $A \in \mathcal{B}(\mathbb{R}^d)$ is bounded with $\lambda^d(A) > 0$, then

$$Q_r(T(A)) = c^r Q_r(A).$$

Proof

Immediate consequence of Lemma 3.2 (or of (6.4)). □

The following lemma shows that the largest quantization error among distributions concentrated on a fixed bounded set appears (asymptotically) for the uniform distribution.

8.2 Lemma
Let $A \in \mathcal{B}(\mathbb{R}^d)$ be bounded with $\lambda^d(A) > 0$. Then

$$\begin{aligned}\max\{Q_r(P) : P(A) = 1\,, P_a \neq 0\} &= Q_r(A)\\ &= Q_r([0,1]^d)\lambda^d(A)^{r/d}.\end{aligned}$$

Proof

Immediate consequence of Hölder's inequality. □

Since one may not expect to be able to find the precise values of $Q_r([0,1]^d)$ for all dimensions d (and all norms), it is of great interest to find bounds. These bounds immediately yield bounds for $Q_r(P)$.

8.1 Ball lower bound

The following lower bound for $Q_r([0,1]^d)$ indicates that the members of an n-optimal partition for a uniform distribution tend to look like a ball. For the l_2-norm it is due to Zador (1963, 1982) and for arbitrary norms this bound is contained in Yamada et al. (1980).

8.3 Proposition

$$Q_r([0,1]^d) \geq M_r(B(0,1)) = \frac{d}{(d+r)\lambda^d(B(0,1))^{r/d}}.$$

Proof

By (6.2) and Theorem 4.16

$$Q_r([0,1]^d) = \inf_{n\geq 1} n^{r/d} M_{n,r}([0,1]^d) \geq M_r(B(0,1)).$$

For the formula for $M_r(B(0,1))$ see Lemma 2.9 □

8.2 Space-filling figures, regular quantizers and upper bounds

For low dimensions good upper bounds for $Q_r([0,1]^d)$ can be obtained by the normalized r-th moment of space-filling figures in $\mathbb{R}^d$. Here a set $A \subset \mathbb{R}^d$ is called **space-filling** if A is compact with $\lambda^d(A) > 0$ and there is a countable family $\mathcal{T}$ of bijective isometries on $\mathbb{R}^d$ such that $\{T(A) : T \in \mathcal{T}\}$ is a tesselation of $\mathbb{R}^d$. This notion depends on the underlying norm. In case the isometries $T \in \mathcal{T}$ can be chosen of the form $T(x) = x + t$, $t \in \mathbb{R}^d$, then A is called **space-filling by translation**. Note that if $S: \mathbb{R}^d \to \mathbb{R}^d$ is a similarity transformation and A is space-filling (by translation), then $S(A)$ is also space-filling (by translation).

Let us mention some properties of space-filling sets.

8.4 Lemma
Let $A \subset \mathbb{R}^d$ be space-filling and let $\mathcal{T}$ be a correponding family of bijective isometries. Choose $a \in C_r(U(A))$ and let $\alpha_A = \{T(a) : T \in \mathcal{T}\}$.

(a) *$\{T(A) : T \in \mathcal{T}\}$ is a locally finite tesselation of $\mathbb{R}^d$.*

(b) *α_A is locally finite.*

(c) *$\lambda^d(\partial A) = 0$ and $int(A) \neq \emptyset$.*

Proof

Set $u = \text{diam}(A)$.

(a) Let $I = \{T \in \mathcal{T} : T(A) \cap B(0,s) \neq 0\}$ for some $s > 0$. Then

$$\bigcup_{T \in I} T(A) \subset B(0, s+u).$$

Therefore

$$|I|\lambda^d(A) = \sum_{T \in I} \lambda^d(T(A)) = \lambda^d(\bigcup_{T \in I} T(A)) \leq \lambda^d(B(0, s+u)) < \infty$$

which implies $|I| < \infty$.

(b) Let $s > 0$ and let $I = \{T \in \mathcal{T} : T(A) \cap B(0, s+u) \neq \emptyset\}$. By (a), I is finite. Let $T \in \mathcal{T} \setminus I$. Then $||y|| > s + u$ for every $y \in T(A)$. Since $T(a) \in C_r(U(T(A)))$ by Lemma 2.1, it follows from Lemma 2.6 (b) that there exists $x \in T(A)$ such that $||x - T(a)|| \leq u$. This gives

$$||T(a)|| \geq ||x|| - ||x - T(a)|| > s + u - u = s.$$

Hence $\{T \in \mathcal{T} : T(a) \in B(0,s)\} \subset I$. This implies that $\alpha_A \cap B(0,s)$ is finite.

(c) Let $S \in \mathcal{T}$. Then $\{S^{-1}T(A) : T \in \mathcal{T}\}$ is a tesselation of $\mathbb{R}^d$. We have

$$\partial A \subset \bigcup_{\substack{T \in \mathcal{T} \\ T \neq S}} (S^{-1}T(A) \cap A)$$

and thus

$$\lambda^d(\partial A) \leq \sum_{\substack{T \in \mathcal{T} \\ T \neq S}} \lambda^d(S^{-1}T(A) \cap A) = 0.$$

This implies $\text{int}(A) \neq \emptyset$. □

Quantizers for $U([0,1]^d)$ can be built with space-filling sets A as follows. For $n \in \mathbb{N}$, consider a rescaled version cA of A, $c = c(n) > 0$, such that $\lambda^d(cA) = 1/n$. Observe that $\{cT(A) : T \in \mathcal{T}\}$ is a tesselation of $\mathbb{R}^d$ consisting of isometric copies of cA. In fact, $T = L + T(0)$, where $L = T - T(0)$ is a bijective isometry with $L(0) = 0$ and hence, L is linear. The bijective isometry $\tilde{T} = L + cT(0)$ then satisfies $\tilde{T}(cA) = cT(A)$. Let

$$I = I(n) = \{T \in \mathcal{T} : cT(A) \subset [0,1]^d\}$$

and $m = m(n) = |I|$. Then $m \leq n$. We thus obtain a tesselation of the cube $[0,1]^d$ consisting of m isometric copies $cT(A)$, $T \in I$, of cA and a region D near the boundary,

$$D = D(n) = [0,1]^d \setminus \bigcup_{T \in I} cT(A);$$

see Figure 8.1. Choose $a \in C_r(U(A))$. Define a **regular n-quantizer** $f_{n,A}$ with respect to A by

$$f_{n,A} = \sum_{T \in I} cT(a) 1_{B_T \cup ((\bigcup_{T \in I} B_T)^c \cap \tilde{B}_T)}, \tag{8.1}$$

provided $I \neq \emptyset$, where $\{B_T : T \in I\}$ is a Borel measurable partition of $\bigcup_{T \in I} cT(A)$ with $B_T \subset cT(A)$ for every $T \in I$ and $\{\tilde{B}_T : T \in I\}$ is a Voronoi partition of $\mathbb{R}^d$ with respect to $\{cT(a) : T \in I\}$.

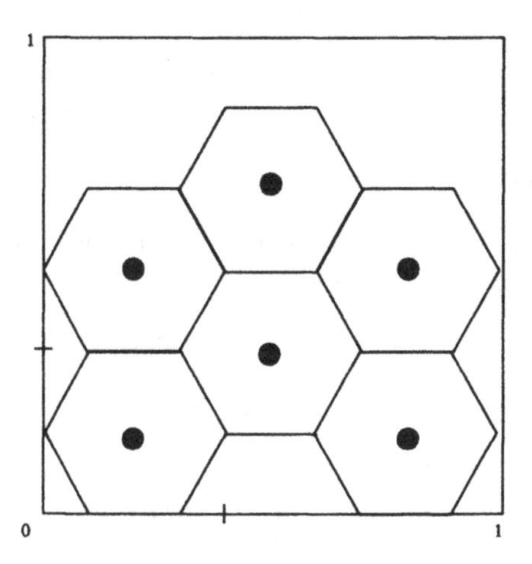

Figure 8.1: Tesselation of $[0,1]^2$ into $m = 6$ regular hexagons and a boundary region, $n = 10$

8.5 Theorem
Let $A \subset \mathbb{R}^d$ be space-filling. Then

$$\lim_{n \to \infty} n^{r/d} \int ||x - f_{n,A}(x)||^r \, dU([0,1]^d)(x) = M_r(A).$$

In particular

$$Q_r([0,1]^d) \leq M_r(A).$$

Proof

By Lemma 2.1, $cT(a) \in C_r(U(cT(A)))$ and $M_r(cT(A)) = M_r(A)$. Therefore

$$\begin{aligned}
&\int \|x - f_{n,A}(x)\|^r \, dU([0,1]^d)(x) \\
&= \sum_{T \in I} \int_{cT(A)} \|x - cT(a)\|^r \, dx + \int_D \min_{T \in I} \|x - cT(a)\|^r \, dx \\
&= n^{-r/d} \frac{m}{n} M_r(A) + \int_D \min_{T \in I} \|x - cT(a))\|^r \, dx \\
&= n^{-r/d} \frac{m}{n} M_r(A) + \int_D \min_{T \in I} \|x - cT(a)\|^r \, dx.
\end{aligned}$$

Since A is space-filling, we have $\lambda^d(D) = \lambda^d(D(n)) = 1 - m(n)/n \to 0$. Let u denote the diameter of A. Then cu is equal to the diameter of $cT(A)$ and

$$cu = c(n)u = n^{-1/d} \lambda^d(A)^{-1/d} u = 0(n^{-1/d}).$$

There exists a constant $\gamma > 1$ such that for every $x \in [0,1]^d$ and sufficiently small $s > 0$ one can find $y \in B(x, \gamma s)$ satisfying

$$B(y, s) \subset B(x, \gamma s) \cap [0,1]^d.$$

This is clearly true for the l_∞-norm and hence for any norm. For $x \in D$ and n large we deduce

$$B(x, \gamma cu) \cap \Big(\bigcup_{T \in I} cT(A) \Big) \neq \emptyset.$$

Otherwise, choose $y \in B(x, \gamma cu)$ such that $B(y, cu) \subset B(x, \gamma cu) \cap [0,1]^d$ and then choose $S \in \mathcal{T}$ such that $y \in cS(A)$. One obtains

$$cS(A) \subset B(y, cu) \subset D,$$

a contradiction. In view of Lemma 2.6 (b) it follows that

$$\min_{T \in I} \|x - cT(a)\| \leq (\gamma + 1) cu, \ n \text{ large}.$$

Therefore

$$\int_D \min_{T \in I} \|x - cT(a)\|^r \, dx \leq ((\gamma + 1) cu)^r \lambda^d(D) = o(n^{-r/d}).$$

This implies the assertion. □

Let the **r-th regular quantization coefficient** of $[0,1]^d$ be defined by

(8.2) $$Q_r^{(R)}([0,1]^d) = \inf\{M_r(A) : A \subset \mathbb{R}^d \text{ space-filling}\}.$$

By the preceding proposition, we get

(8.3) $$Q_r([0,1]^d) \le Q_r^{(R)}([0,1]^d).$$

A basic question is whether equality holds in (8.3). The regular quantizer problem consists in finding a space-filler $A \subset \mathbb{R}^d$ such that $Q_r^{(R)}([0,1]^d) = M_r(A)$. Both problems are unsolved for $d \ge 3$.

One technique for obtaining upper bounds is to select space-fillers in higher dimensional spaces by forming products of two (or more) lower dimensional space-fillers.

8.6 Lemma
Let the underlying norm be the l_r-norm, $1 \le r < \infty$. Let $A \subset \mathbb{R}^d$ and $B \subset \mathbb{R}^k$ be space-filling. Then $A \times B \subset \mathbb{R}^{d+k}$ is space-filling and

$$M_r(A \times B) = \frac{M_r(A)\lambda^d(A)^{r/d} + M_r(B)\lambda^k(B)^{r/k}}{\lambda^d(A)^{r/(d+k)}\lambda^k(B)^{r/(d+k)}}.$$

Proof

Clearly $A \times B$ is space-filling in $\mathbb{R}^{d+k}$. Furthermore,

$$\begin{aligned} V_r(U(A \times B)) &= V_r(U(A)) + V_r(U(B)) \\ &= M_r(A)\lambda^d(A)^{r/d} + M_r(B)\lambda^k(B)^{r/k}. \end{aligned}$$

This implies the assertion. □

8.3 Lattice quantizers

The Voronoi regions of lattices in $\mathbb{R}^d$ provide an interesting class of regular quantizers.

A (d-dimensional) lattice in $\mathbb{R}^d$ is a locally finite additive subgroup of $\mathbb{R}^d$ which spans $\mathbb{R}^d$. Equivalently, a lattice is a subset of the form $\Lambda = \mathbb{Z}y_1 + \ldots + \mathbb{Z}y_d$ for some (vector space) basis $\{y_1, \ldots, y_d\}$ of $\mathbb{R}^d$; such a basis is called basis of Λ. The volume of a fundamental parallelotope $\left\{\sum_{i=1}^d t_i y_i : 0 \le t_i \le 1 \text{ for } 1 \le i \le d\right\}$ of a lattice Λ does not depend on the choice of the basis $\{y_1, \ldots, y_d\}$ of Λ and is denoted by $\det(\Lambda)$. Then $\det(\Lambda)$ is the absolute value of the determinant of the matrix with rows $y_1, \ldots, y_d$. Fundamental parallelotopes of Λ are space-filling by translation with Λ as set of translation vectors and any space-filler $A \subset \mathbb{R}^d$ of this type satisfies $\lambda^d(A) = \det(\Lambda)$.

Now consider the Voronoi diagram $\{W(a|\Lambda) : a \in \Lambda\}$ of a lattice $\Lambda \subset \mathbb{R}^d$. Obviously

$$W(a|\Lambda) = W(0|\Lambda) + a, \; a \in \Lambda.$$

If the Voronoi diagram of Λ is a tesselation of $\mathbb{R}^d$ we say that Λ is **admissible**. For strictly convex norms every lattice is admissible (see (1.7) and Theorem 1.5).

8.7 Lemma
Let $\Lambda \subset \mathbb{R}^d$ be a lattice.

(a) *$W(0|\Lambda)$ is compact and $\lambda^d(W(0|\Lambda)) \geq \det(\Lambda)$.*

(b) *Λ is admissible if and only if $\lambda^d(W(0|\Lambda)) = \det(\Lambda)$ and $\lambda^d(\partial W(0|\Lambda)) = 0$.*

Proof

(a) Choose $s > 0$ such that $B(0, s)$ contains a fundamental parallelotope of Λ. Then $\{B(a, s) : a \in \Lambda\}$ is a covering of $\mathbb{R}^d$. This implies $W(0|\Lambda) \subset B(0, s)$ and hence, $W(0|\Lambda)$ is compact. Furthermore, if B denotes a fundamental parallelotope of Λ, then

$$\begin{aligned}\lambda^d(W) &= \sum_{a\in\Lambda} \lambda^d(W \cap (B + a)) \\ &= \sum_{a\in\Lambda} \lambda^d((W - a) \cap B) \\ &\geq \lambda^d(B) = \det(\Lambda),\end{aligned}$$

where $W = W(0|\Lambda)$. Here the inequality follows from the fact that the Voronoi diagram of Λ is a covering if $\mathbb{R}^d$; see Proposition 1.1.

(b) If Λ is admissible, then in view of (a), $W(0|\Lambda)$ is space-filling by translations with Λ as set of translation vectors. This implies $\lambda^d(W(0|\Lambda)) = \det(\Lambda)$. If Λ is not admissible and $\lambda^d(\partial W(0|\Lambda)) = 0$, then

$$\operatorname{int} W(a_1|\Lambda) \cap \operatorname{int} W(a_2|\Lambda) \neq \emptyset$$

for some $a_1, a_2 \in \Lambda$, $a_1 \neq a_2$. Hence, there exist $x_1, x_2 \in \operatorname{int} W(0|\Lambda)$, $x_1 \neq x_2$, such that $x_1 - x_2 \in \Lambda$. Let $b = x_1 - x_2$. Choose $\varepsilon > 0$ such that $B(x_i, \varepsilon) \subset W(0|\Lambda)$ and $B(x_1, \varepsilon) \cap B(x_2, \varepsilon) = \emptyset$. Set $A = W(0|\Lambda) \setminus B(x_1, \varepsilon)$. Then

$$B(x_1, \varepsilon) = B(x_2, \varepsilon) + b \subset A + b$$

which yields

$$W(0|\Lambda) \subset A \cup (A + b).$$

This implies that $\{A + a : a \in \Lambda\}$ is a covering of $\mathbb{R}^d$, hence $\lambda^d(A) \geq \det(\Lambda)$. We obtain

$$\lambda^d(W(0|\Lambda)) > \lambda^d(A) \geq \det(\Lambda).$$

□

It is remarkable that by part (b) of the preceding lemma, the volume of $W(0|\Lambda)$ does not depend on the underlying norm as long as admissibility holds.

There are lattices which are not admissible. This is illustrated by the following example.

8.8 Example
Let the underlying norm on $\mathbb{R}^2$ be the l_1-norm and let $\Lambda = \mathbb{Z}(-1,1) + \mathbb{Z}(4,0)$. Then $\Lambda = \{a \in \mathbb{Z}^2 : a_1 + a_2 \in 4\mathbb{Z}\}$, $\det(\Lambda) = 4$,

$$\begin{aligned} W(0|\Lambda) = &\{x \in \mathbb{R}^2 : |x_1| + |_2| \leq 1\} \\ &\cup ([-2,0]^2 \cap \{x \in \mathbb{R}^2 : x_1 + x_2 \geq -2\}) \\ &\cup ([0,2]^2 \cap \{x \in \mathbb{R}^2 : x_1 + x_2 \leq 2\}) \end{aligned}$$

and $\lambda^2(W(0|\Lambda)) = 5$; see Figure 8.2. It follows from Lemma 8.7 that Λ is not admissible (for the l_1-norm).

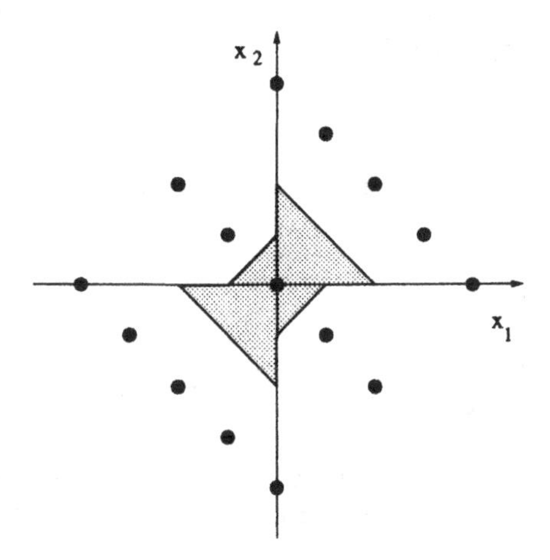

Figure 8.2: Voronoi region $W(0|\Lambda)$ with respect to the l_1-norm for a nonadmissible lattice

As concerns the convexity of $W(0|\Lambda)$ one can modify Remark 1.9 as follows: if $W(0|\Lambda)$ is convex for every lattice $\Lambda \subset \mathbb{R}^d$, then the underlying norm is euclidean (cf. Gruber, 1974, Theorem 2).

If $\Lambda \subset \mathbb{R}^d$ is an admissible lattice, then we know from Lemma 8.7 that the Voronoi region $W(0|\Lambda)$ is space-filling by translation with Λ as set of translation vectors. Thus Therorem 8.5 applies to the n-quantizer $f_{n,\Lambda} = f_{n,W(0|\Lambda)}$ for $U([0,1]^d)$. Note that $W(0|\Lambda)$ is symmetric (about the origin) and hence

$$M_r(W(0|\Lambda)) = \frac{\int_{W(0|\Lambda)} \|x\|^r \, dx}{\det(\Lambda)^{(d+r)/d}}; \tag{8.4}$$

cf. Example 2.3. For $n \in \mathbb{N}$, let

$$\begin{aligned} \alpha_{n,\Lambda} &= \{ca : a \in \Lambda, cW(a|\Lambda) \subset [0,1]^d\}, \\ c &= c(n) = (n \det(\Lambda))^{-1/d}. \end{aligned} \tag{8.5}$$

8.9 Theorem
Let $\Lambda \subset \mathbb{R}^d$ be an admissible lattice and let X be $U([0,1]^d)$-distributed. Then

$$\lim_{n\to\infty} n^{r/d} \int \min_{b\in\alpha_{n,\Lambda}} \|x-b\|^r \, dU([0,1]^d)(x) = M_r(W(0|\Lambda))$$

and

$$n^{1/d} \min_{b\in\alpha_{n,\Lambda}} \|X-b\| \xrightarrow{D} \mu \text{ as } n\to\infty,$$

where μ is a probability on $\mathbb{R}_+$ with distribution function

$$F_\mu(t) = \lambda^d\big((\det(\Lambda)^{-1/d} W(0|\Lambda)) \cap B(0,t)\big).$$

In particular

$$Q_r([0,1]^d) \le M_r(W(0|\Lambda)).$$

Proof

We have

$$E \min_{b\in\alpha_{n,\Lambda}} \|X-b\|^r = E\|X - f_{n,\Lambda}\|^r$$

(when $f_{n,\Lambda}$ is defined in (8.1) with respect to the center $a = 0$). So the first assertion follows from Theorem 8.5. Now observe that $\alpha_{n,\Lambda}$ does not depend on r and

$$M_r(W(0|\Lambda)) = \int z^r \, d\mu(z).$$

Since $\text{supp}(\mu)$ is compact, the convergence of moments,

$$\lim_{n\to\infty} n^{r/d} E \min_{b\in\alpha_{n,\Lambda}} \|X-b\|^r = \int z^r \, d\mu(z) \text{ for every } 1 \le r < \infty,$$

implies the desired distributional convergence (cf. Hoffmann-Jørgensen, 1994, 5.13). □

Let the **r-th lattice quantization coefficient** of $[0,1]^d$ be defined by

(8.6) $$Q_r^{(L)}([0,1]^d) = \inf\{M_r(W(0|\Lambda)) : \Lambda \subset \mathbb{R}^d \text{ admissible lattice}\}.$$

Then

(8.7) $$Q_r([0,1]^d) \le Q_r^{(R)}([0,1]^d) \le Q_r^{(L)}([0,1]^d).$$

The lattice quantizer problem consists in finding an admissible lattice Λ such that $Q_r^{(L)}([0,1]^d) = M_r(W(0|\Lambda))$.

8.10 Remark

(a) Suppose $A \subset \mathbb{R}^d$ is space-filling by translation, where the corresponding set of translation vectors is an admissible lattice Λ. Then

$$M_r(W(0|\Lambda)) \leq M_r(A).$$

In fact, since

$$E \min_{b \in \alpha_{n,\Lambda}} \|X - b\|^r \leq E\|X - f_{n,A}\|^r,$$

where X is $U([0,1]^d)$-distributed, the above inequality follows from Theorems 8.5 and 8.9.

(b) Suppose $A \subset \mathbb{R}^d$ is a convex space-filler by translation. Then A is a centrally symmetric polytope (i. e. $-(A - x) = A - x$ for some $x \in \mathbb{R}^d$) and admits as set of translation vectors a lattice Λ (cf. McMullen, 1980). By (a) we have

$$M_r(W(0|\Lambda)) \leq M_r(A)$$

provided Λ is admissible. Thus we obtain

$$Q_r^{(L)}([0,1]^d) = \inf\{M_r(A) : A \subset \mathbb{R}^d \text{ space-filling polytope by translation}\}$$

for euclidean norms.

(c) Suppose the ball $B(0,1)$ is space-filling. Then the ball is obviously space-filling by translation. By (b), there exists a lattice Λ as set of translation vectors. Then $W(0|\Lambda) = B(0,1)$ and hence Λ is admissible. In view of Proposition 8.3 we obtain

$$\begin{aligned} M_r(B(0,1)) &= Q_r([0,1]^d) \\ &= Q_r^{(R)}([0,1]^d) = Q_r^{(L)}([0,1]^d). \end{aligned}$$

Conversely, if $M_r(B(0,1)) = Q_r^{(L)}([0,1]^d)$ holds (for some r) and if the lattice quantizer problem has a solution Λ, then $B(0,1)$ is space-filling. To see this, choose $s > 0$ such that $\lambda^d(B(0,s)) = \det(\Lambda)$. By (the proof of) Lemma 2.9, we have $B(0,s) \subset W(0|\Lambda)$. To verify the converse inclusion, assume that there exists $x \in W(0,\Lambda)$ with $s < \|x\|$. Since the distance function $d(\cdot,\Lambda)$ is continuous on $\mathbb{R}^d$, one obtains $B(x,\varepsilon) \subset \left(\bigcup_{a \in \Lambda} B(a,s)\right)^c$ for some $\varepsilon > 0$. Choose $a \in \Lambda$ such that $\lambda^d(B(x,\varepsilon) \cap W(a|\Lambda)) > 0$. Then

$$\begin{aligned} \lambda^d(B(0,s)) &< \lambda^d(B(a,s)) + \lambda^d(B(x,\varepsilon) \cap W(a|\Lambda)) \\ &\leq \lambda^d(W(a|\Lambda)) = \det(\Lambda), \end{aligned}$$

a contradiction. Hence $B(0,s) = W(0|\Lambda)$ and so $B(0,1)$ is space-filling.

The lattices in the following examples are related to optimality results.

8.11 Example (Standard lattice $\mathbb{Z}^d$)
Let $\Lambda = \mathbb{Z}^d$ and let the underlying norm be the l_p-norm, $1 \leq p \leq \infty$. Then $\det(\Lambda) = 1$ and $W(0|\Lambda) = [-\frac{1}{2}, \frac{1}{2}]^d$. In particular, Λ is admissible. For computations of the normalized r-th moments of $W(0|\Lambda)$ see Example 4.17. In case $p = \infty$, the limiting measure μ in Theorem 8.9 is given by

$$F_\mu(t) = (2t)^d, \; 0 \leq t \leq 1/2.$$

For $d = 1$, one obtains $\mu = U([0, \frac{1}{2}])$.

8.12 Example (Hexagonal lattice in $\mathbb{R}^2$)
Let $d = 2$ and let $\Lambda = \mathbb{Z}(1,0) + \mathbb{Z}(1/2, \sqrt{3}/2)$. Here we have $\det(\Lambda) = \sqrt{3}/2$. If the underlying norm is the l_2-norm, then $W(0|\Lambda)$ is a regular hexagon,

$$\begin{aligned} W(0|\Lambda) &= \{x \in \mathbb{R}^2 : |x_1| \leq 1/2, |x_1| + \sqrt{3}|x_2| \leq 1\} \\ &= \operatorname{conv}\{\pm(\frac{1}{2}, \frac{1}{2\sqrt{3}}), \pm(0, \frac{1}{\sqrt{3}}), \pm(\frac{1}{2}, -\frac{1}{2\sqrt{3}})\}; \end{aligned}$$

see Figure 8.3.

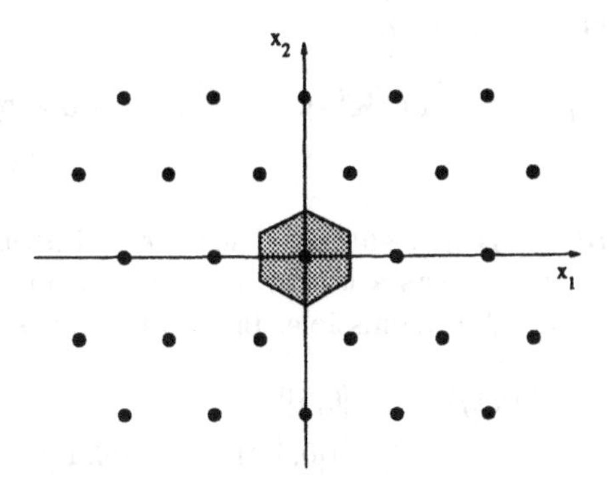

Figure 8.3: Voronoi region $W(0|\Lambda)$ with respect to the l_2-norm for the hexagonal lattice

We have

$$M_r(W(0|\Lambda)) = \frac{8 \cdot 2^{r/2}}{3^{(2+r)/4}} \int_0^{1/2} \int_0^{(1-x_1)/\sqrt{3}} (x_1^2 + x_2^2)^{r/2} dx_2 dx_1.$$

In case $r = 1$ and $r = 2$ one obtains

$$M_1(W(0|\Lambda)) = \frac{2 + 3\log(\sqrt{3})}{3^{7/4}\sqrt{2}} = 0.37771\ldots$$

and

$$M_2(W(0|\Lambda)) = \frac{5}{18\sqrt{3}} = 0.1603....$$

If the underlying norm is the l_1-norm, then $W(0|\Lambda)$ is the (nonregular) hexagon

$$\begin{aligned} W(0|\Lambda) &= \{x \in \mathbb{R}^2 : |x_1| \le 1/2,\ |x_1| + |x_2| \le (1+\sqrt{3})/4\} \\ &= \operatorname{conv}\{\pm(\frac{1}{2}, \frac{\sqrt{3}-1}{4}), \pm(0, \frac{1+\sqrt{3}}{4}), \pm(\frac{1}{2}, \frac{1-\sqrt{3}}{4})\}. \end{aligned}$$

Since $\lambda^2(W(0|\Lambda)) = \sqrt{3}/2$, Λ is admissible by Lemma 8.7.

8.13 Example (Lattices D_d)

Let $\Lambda = \{a \in \mathbb{Z}^d : \sum_{i=1}^d a_i \text{ even}\}$. This lattice is usually called of type D_d or checker-board lattice. Here we have $\det(\Lambda) = 2$. If the underlying norm is the l_2-norm, we obtain

$$\begin{aligned} W(0|\Lambda) &= [-1,1],\ d=1, \\ W(0|\Lambda) &= \{x \in \mathbb{R}^d : \sum_{i \in I} |x_i| \le 1 \text{ for every} \\ &\qquad I \subset \{1, \ldots, d\}, |I| = 2\},\ d \ge 2. \end{aligned}$$

In case $d = 2$, $W(0|\Lambda)$ is the unit l_1-ball and in dimension $d = 3$, $W(0|\Lambda)$ is a rhombic dodecahedron. Let $r = 2$ and write $W = W(0|\Lambda)$. Since W is invariant under permutations of the coordinates, one gets

$$\begin{aligned} \int_W \|x\|^2\,dx &= d \int_W x_1^2\,dx \\ &= d \int_{-1}^{1} x_1^2 \lambda^{d-1}(W_{x_1})\,dx_1, \end{aligned}$$

where W_{x_1} denotes the x_1-section of W,

$$\begin{aligned} W_{x_1} = \{y \in \mathbb{R}^{d-1} &: \sum_{i \in I} |y_i| \le 1 \text{ for every } I \subset \{1, \ldots, d-1\}, \\ &|I| = 2,\ |y_i| \le 1 - |x_1| \text{ for } 1 \le i \le d-1\},\ |x_1| \le 1. \end{aligned}$$

Using

$$\begin{aligned} \lambda^{d-1}(W_{x_1}) &= 2 - 2^{d-1}|x_1|^{d-1},\ |x_1| \le \frac{1}{2}, \\ \lambda^{d-1}(W_{x_1}) &= 2^{d-1}(1 - |x_1|)^{d-1},\ \frac{1}{2} \le |x_1| \le 1 \end{aligned}$$

we deduce

$$\int_W \|x\|^2\,dx = \frac{d}{6} + \frac{1}{d+1}.$$

This implies the formula

$$M_2(W(0|\Lambda)) = \frac{1}{2^{2/d}}(\frac{d}{12} + \frac{1}{2(d+1)}),\ d \geq 1$$

(cf. Conway and Sloane, 1993, p. 462). In particular

$$M_2(W(0|\Lambda)) = \frac{3}{4^{1/3}8} = 0.2362\ldots,\ d = 3,$$

$$M_2(W(0|\Lambda)) = \frac{13}{2^{1/2}30} = 0.3064\ldots,\ d = 4.$$

If the underlying norm is the l_1-norm, then $W(0|\Lambda)$ coincides with the above l_2-Voronoi region and Λ is thus admissible. Here we obtain for $r = 1$

$$M_1(W(0|\Lambda)) = \frac{1}{2^{1/d}}(\frac{d}{4} + \frac{1}{2(d+1)}),\ d \geq 1.$$

In particular,

$$M_1(W(0|\Lambda)) = \frac{7}{2^{1/3}8} = 0.6944\ldots,\ d = 3.$$

8.14 Example (Dual lattices D_d^*)
The dual lattice of the lattice D_d is defined by

$$D_d^* = \left\{x \in \mathbb{R}^d : \sum_{i=1}^d a_i x_i \in \mathbb{Z} \text{ for every } a \in D_d\right\}.$$

Then

$$D_d^* = \mathbb{Z}^d + \mathbb{Z}(\frac{1}{2}, \ldots, \frac{1}{2})$$

and

$$2D_d^* = (2\mathbb{Z})^d \cup (2\mathbb{Z}+1)^d$$

Note that $2D_1^* = \mathbb{Z}$ and $2D_2^* = D_2$. Let $\Lambda = 2D_d^*$. Then $\det(\Lambda) = 2^d/\det(D_d) = 2^{d-1}$ If the underlying norm is the l_2-norm, it is not difficult to verify that

$$W(0|\Lambda) = \left\{x \in \mathbb{R}^d : \sum_{i=1}^d |x_i| \leq \frac{d}{2}\right\} \cap [-1,1]^d.$$

For $d = 3$, this Voronoi region is a truncated octahedron; see Figure 8.4. It is more difficult to compute the normalized second moment of $W(0|\Lambda)$. We obtain

$$M_2(W(0|\Lambda)) = \frac{19}{2^{1/3}64} = 0.2356\ldots,\ d = 3.$$

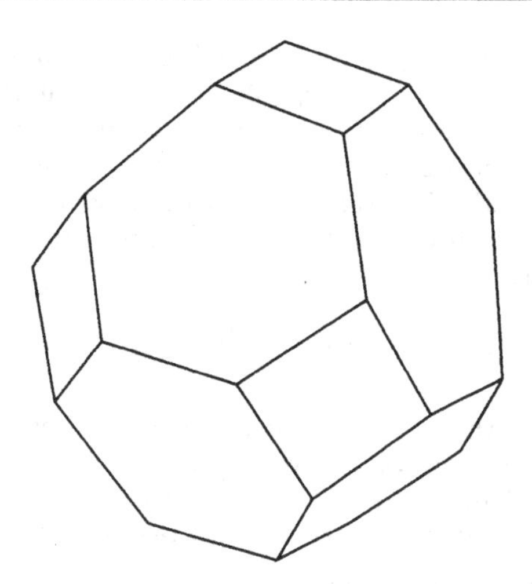

Figure 8.4: Truncated octahedron

Note that this moment is slightly smaller than $M_2(W(0|D_3))$. For $d = 4$, Λ and D_4 are similar. In fact, the similarity transformation

$$T : \mathbb{R}^4 \to \mathbb{R}^4, \quad T(x) = (x_1 + x_2, x_1 - x_2, x_3 + x_4, x_3 - x_4)$$

with scaling factor $\sqrt{2}$ satisfies $T(D_4) = \Lambda$. Therefore, $W(0|\Lambda) = TW(0|D_4)$ which yields

$$M_2(W(0|\Lambda)) = M_2(W(0|D_4)) = \frac{13}{2^{1/2}30} = 0.3064\ldots, \; d = 4$$

A general formula for $M_2(W(0|\Lambda))$ can be found in Conway and Sloane, 1993, pp. 470-471. The above upper bounds for $Q_r([0,1]^d)$, $d = 3, 4$, are close to the ball lower bounds given in Proposition 8.3. We have

$$M_2(B(0,1)) = \left(\frac{3}{4\pi}\right)^{2/3}\frac{3}{5} = 0.2309\ldots, \; d = 3,$$

$$M_2(B(0,1)) = \frac{2\sqrt{2}}{3\pi} = 0.3001\ldots, \; d = 4.$$

If the underlying norm is the l_1-norm, the Voronoi region generated by 0 does not change and so Λ is admissible. For $d = 3$ and $r = 1$, we obtain

$$M_1(W(0|\Lambda)) = \frac{35}{4^{1/3}32} = 0.6890\ldots, d = 3.$$

This moment is smaller than $M_1(W(0|D_3))$. The corresponding ball lower bound is given by

$$M_1(B(0,1)) = \left(\frac{3}{4}\right)^{4/3} = 0.6814\ldots, \; d = 3.$$

If the underlying norm is the l_2-norm, $Q_r([0,1]^d)$ is only known for $d = 1$ and $d = 2$. We will see below that for $d = 2$, $Q_r([0,1]^2) = M_r(W(0|\Lambda))$ holds, where Λ is the hexagonal lattice described in Example 8.12. In particular, Λ solves the lattice quantizer problem. For dimension $d = 3$, the Example 8.14 shows that the normalized second moment of the lattice D_3^* (truncated octahedron) is very close to the ball lower bound. It is known that D_3^* is a solution of the lattice quantizer problem for $r = 2$ (cf. Barnes and Sloane, 1983), so

$$(8.8) \qquad Q_2^{(L)}([0,1]^3) = M_2(W(0|D_3^*)) = \frac{19}{2^{1/3}64} = 0.2356\ldots, \; l_2\text{-norm.}$$

For $d \geq 4$, solutions of the lattice quantizer problem are not known. Conway and Sloane (1993) give a comprehensive survey of the best known lattice quantizers for $r = 2$ among them D_4^* (or D_4) and D_5^*. For recent improvements see Agrell and Eriksson (1998). (Note that these authors present the value of $M_2(A)/d$ for $A \subset \mathbb{R}^d$.)

If the underlying norm is the l_1-norm, the Example 8.14 shows that the normalized first moment of D_3^* is close to the ball lower bound and hence, D_3^* provides a good quantizer for $U([0,1]^3)$ in case $r = 1$. However, optimality results are not known for $d \geq 3$. A trivial case occurs for $d = 2$, where the ball $B(0,1) = W(0|D_2)$ is space-filling. Therefore

$$(8.9) \qquad Q_r([0,1]^2) = M_r(B(0,1)) = \frac{2}{(2+r)2^{r/2}}, \; l_1\text{-norm.}$$

A further trivial case concerns the l_∞-norm. Here $B(0,1) = W(0|\mathbb{Z}^d)$ is again space-filling and so

$$(8.10) \qquad Q_r([0,1]^d) = M_r(B(0,1)) = \frac{d}{(d+r)2^r}, \; l_\infty\text{-norm}$$

(cf. Example 4.17).

The following result is due to Fejes Tóth (1959, 1972).

8.15 Theorem
Let $d = 2$ and suppose the underlying norm is the l_2-norm. Let Λ be the hexagonal lattice. Then

$$Q_r([0,1]^2) = M_r(W(0|\Lambda)).$$

Proof

Set $A = W(0|\Lambda)$ and recall that A is a regular hexagon. For every $n \in \mathbb{N}$ and every $\alpha \subset A$ with $|\alpha| = n$, we have

$$\int_A \min_{a\in\alpha} \|x-a\|^r \, dx \geq n \int_{n^{-1/2}A} \|x\|^r \, dx$$

(cf. Fejes Tóth, 1972, p. 81). Let $\alpha \in C_{n,r}(U(A))$. It follows from Theorem 4.1 and Lemma 2.6 (a) that $\alpha \subset A$. Therefore

$$\begin{aligned}\det(\Lambda)V_{n,r}(U(A)) &= \int_A \min_{a\in\alpha} \|x-a\|^r \, dx \\ &\geq n \int_{n^{-1/2}A} \|x\|^r \, dx \\ &= n^{-r/2} M_r(A) \det(\Lambda)^{(2+r)/2}\end{aligned}$$

which yields

$$n^{r/2} V_{n,r}(U(A)) \geq M_r(A) \; \det(\Lambda)^{r/2}, \; n \in \mathbb{N}.$$

This implies

$$Q_r(A) \geq M_r(A) \det(\Lambda)^{r/2}$$

and hence

$$Q_r([0,1]^d) \geq M_r(A).$$

This together with Theorem 8.9 gives the assertion. □

8.16 Remark

The above results allow to prove by a quantization argument that $\alpha_{n,\Lambda}$ is uniformly distributed in $[0,1]^d$ for every admissible lattice $\Lambda \subset \mathbb{R}^d$ with convex Voronoi region $W(0|\Lambda)$ in the sense that

$$\frac{1}{|\alpha_{n,\Lambda}|} \sum_{b\in\alpha_{n,\Lambda}} \delta_b \xrightarrow{D} U([0,1]^d) \text{ as } n \to \infty.$$

First, observe that $A = W(0|\Lambda)$ is the unit ball of some norm $\| \; \|_0$. Then forget the underlying norm which was only used to form the Voronoi region $W(0|\Lambda)$ and through this $\alpha_{n,\Lambda}$ and proceed with the norm $\| \; \|_0$. The Voronoi region $W(0|\Lambda, \| \; \|_0)$ with respect to $\| \; \|_0$ coincides with A. Therefore, by Theorem 8.9 and Proposition 8.3

$$\lim_{n\to\infty} n^{r/d} \int \min_{b\in\alpha_{n,\Lambda}} \|x-b\|_0^r \, dU([0,1]^d)(x) = M_r(A, \| \; \|_0) = Q_r([0,1]^d, \| \; \|_0).$$

The assertion now follows from Theorem 7.5.

8.4 Quantization coefficients of one-dimensional distributions

In Tables 8.1–8.3 one can find the quantization coefficients

$$Q_r(P) = \frac{1}{(1+r)2^r} \left(\int \left(\frac{dP}{d\lambda} \right)^{1/(1+r)} d\lambda \right)^{1+r}$$

of several univariate absolutely continuous distributions. As an illustration, the Figure 8.5 shows the densities of three hyper-exponential distributions with variance equal to one and small, moderate and large second quantization coefficient, respectively.

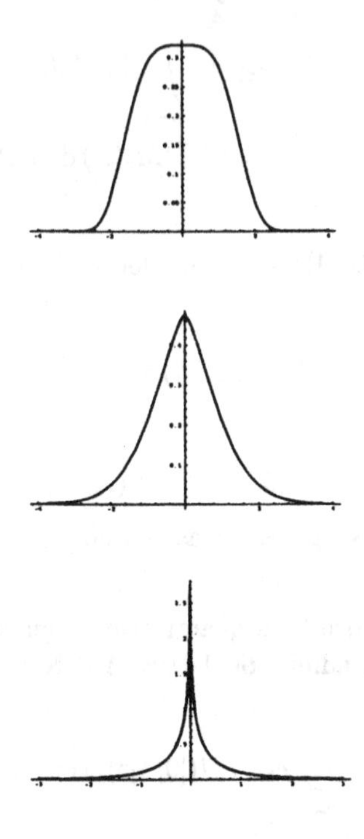

Figure 8.5: Densities of hyper-exponential distributions $P = H(a, b)$ with variance equal to one and $Q_2(P) = 1.8470$ (top), $Q_2(P) = 3.3106$ (center), $Q_2(P) = 8.1000$ (bottom).

P	$Q_r(P)$
Normal $N(0,\sigma^2)$	$(\frac{\sigma^2\pi}{2})^{r/2}(1+r)^{(r-1)/2}$
Logistic $L(a)$	$(\frac{a}{2})^r \frac{\Gamma(\frac{1}{1+r})^{2+2r}}{(1+r)\Gamma(\frac{2}{1+r})^{1+r}}$
Double Exponential $DE(a)$	$(a(1+r))^r$
Double Gamma $D\Gamma(a,b)$	$a^r(1+r)^{b+r-1}\frac{\Gamma(\frac{b+r}{1+r})^{1+r}}{\Gamma(b)}$
Hyper-exponential $HE(a,b)$	$(\frac{a\Gamma(\frac{1}{b})}{b})^r(1+r)^{(1+r-b)/b}$
Uniform $U([a,b])$	$\frac{1}{1+r}(\frac{b-a}{2})^r$
Triangular $T(a,b;c)$	$\frac{2}{(2+r)^{1+r}}(\frac{(1+r)(b-a)}{2})^r$
Exponential $E(a)$	$(\frac{a(1+r)}{2})^r$
Gamma $\Gamma(a,b)$	$(\frac{a}{2})^r(1+r)^{b+r-1}\frac{\Gamma(\frac{b+r}{1+r})^{1+r}}{\Gamma(b)}$
Weibull $W(a,b)$	$(\frac{a(1+r)^{1/b}}{2b})^r\,\Gamma(\frac{b+r}{b(1+r)})^{1+r}$
Pareto $P(a,b), b>r$	$\frac{b}{b-r}(\frac{a(1+r)}{2(b-r)})^r$

Table 8.1: Quantization coefficients

P	$Q_2(P)$
$N(0,1)$	$\frac{\sqrt{3}\,\pi}{2} = 2.7206...$
$L(\sqrt{3}/\pi)$	$\frac{\Gamma(\frac{1}{3})^6}{4\pi^2\Gamma(\frac{2}{3})^3} = 3.7709...$
$DE(1/\sqrt{2})$	$\frac{9}{2} = 4.5$
$D\Gamma(a,b)$ $a^2 = \frac{1}{b(1+b)}$	$\frac{3^{b+1}\Gamma(\frac{b+2}{3})^3}{\Gamma(b+2)}$ range: $(0, 3\Gamma(\frac{2}{3})^3) = (0,\ 7.4488...)$
$HE(a,b)$ $a^2 = \frac{\Gamma(\frac{1}{b})}{\Gamma(\frac{3}{b})}$	$\frac{\Gamma(\frac{1}{b})^3\ 3^{(3-b)/b}}{b^2\Gamma(\frac{3}{b})}$ range: $(1, \infty)$
$U([a,b])$, $b = a + 2\sqrt{3}$	1
$T(a, b; \frac{a+b}{2})$, $b = a + 2\sqrt{6}$	$\frac{27}{16} = 1.6875$
$E(1)$	$\frac{9}{4} = 2.25$
$\Gamma(a,b)$ $a^2 = \frac{1}{b}$	$\frac{3^{b+1}\Gamma(\frac{b+2}{3})^3}{4\Gamma(b+1)}$ range: $(\frac{3\Gamma(\frac{2}{3})^3}{4}, \frac{\sqrt{3}\pi}{2}) = (1.8622..., 2.7206...)$
$W(a,b)$ $a^2 = \frac{1}{\Gamma(\frac{b+2}{b}) - \Gamma(\frac{b+1}{b})^2}$	$\frac{9^{1/b}\Gamma(\frac{b+2}{3b})^3}{4b^2[\Gamma(\frac{b+2}{b} - \Gamma(\frac{b+1}{b})^2]}$ range: $[2.1555, \infty)$
$P(a,b), b > 2$, $a^2 = \frac{(b-2)(b-1)^2}{b}$	$\frac{9}{4}(\frac{b-1}{b-2})^2$ range: $(\frac{9}{4}, \infty)$

Table 8.2: $r = 2$. Quantization coefficients of distributions P with $V_2(P) = 1$

P	$Q_1(P)$
$N(0, \frac{\pi}{2})$	$\frac{\pi}{2} = 1.5707...$
$L(\frac{1}{2\log 2})$	$\frac{\pi^2}{8\log 2} = 1.7798...$
$DE(1)$	2
$D\Gamma(a,b)$ $a = \frac{1}{b}$	$\frac{2^b \Gamma(\frac{b+1}{2})^2}{\Gamma(b+1)}$ range: $(0, \pi)$
$HE(a,b)$ $a = \frac{\Gamma(\frac{1}{b})}{\Gamma(\frac{2}{b})}$	$\frac{\Gamma(\frac{1}{b})^2 2^{(2-b)/b}}{b\Gamma(\frac{2}{b})}$ range : $(1, \infty)$
$U([a,b])$, $b = a+4$	1
$T(a,b;\ \frac{a+b}{2})$, $b = a+6$	$\frac{4}{3} = 1.3333...$
$E(\frac{1}{\log 2})$	$\frac{1}{\log 2} = 1.4426...$
$P(a,b), b > 1$, $a = \frac{b-1}{b(2^{1/b}-1)}$	$\frac{1}{(b-1)(2^{1/b}-1)}$ range: $(\frac{1}{\log 2}, \infty)$

Table 8.3: $r = 1$. Quantization coefficients of distributions P with $V_1(P) = 1$

Notes

The issue of space-filling sets in the quantization setting was raised by Gersho (1979) for the l_2-norm. Expositions concerning tesselations (tilings) can be found in Grünbaum and Shephard (1986) and Schulte (1993). For a background on lattices, we refer to the books of Cassels (1959) and Gruber and Lekkerkerker (1987).

Gersho (1979) contains upper bounds for $Q_r([0,1]^d)$ of the type (8.3) for the l_2-norm. Theorems 8.5 and 8.9 provide a rigorous derivation. The obervation in Theorem 8.9 concerning the distributional convergence seems to be new. Different proofs for the Hexagon-Theorem 8.15 can be found in Newman (1982) (for $r = 2$), Wong (1982) (also for $r = 2$) and Haimovich and Magnati (1988). Discussions and applications of this theorem are contained in Bollobás (1972,1973). The quantization coefficient $Q_1([0,1]^d)$ appears in upper bounds for limiting constants in the euclidean traveling salesman problem; see Goddyn (1990).

According to Theorems 8.9 and 8.15, the hexagon quantizer is asymptotically n–optimal for $P = U([0,1]^2)$ of every order r. This result can be extended to bivariate nonuniform distributions P with a continuous density using a piecewise hexagon quantizer depending on P and r in the spirit of Lemma 7.2 (but now with $m \to \infty$, $m/n \to 0$ and the n_i–optimal quantizer for $P(\cdot|A_i)$ replaced by a hexagon quantizer). See McClure (1980, p. 197) for $r = 2$ with an unpublished proof. Su (1997) showed that this design yields an asymptotically n–optimal quantizer for every r.

8.17 Conjecture
$Q_r([0,1]^3) = M_r(W(0|D_3^*))$ for every $r \in [1,\infty)$ when the underlying norm on $\mathbb{R}^3$ is the l_2–norm (cf. Example 8.14, (8.8), and Remark 10.11(c)).

9 Random quantizers and quantization coefficients

In this section we determine the asymptotics of a stochastic version of the quantization problem and derive further upper bounds and the d-asymptotics for the quantization coefficients.

9.1 Asymptotics for random quantizers

Let X be a $\mathbb{R}^d$-valued random variable with distribution P, let $\|\ \|$ denote any norm on $\mathbb{R}^d$, and let $1 \leq r < \infty$.

9.1 Theorem
Suppose $P = P_a$. Let $Y_1, Y_2, \ldots$ be i.i.d. $\mathbb{R}^d$-valued random variables with distribution Q independent of X and let $g = dQ_a/d\lambda^d$.

(a) *Assume $P(g > 0) = 1$. Then*

$$n^{1/d} \min_{1 \leq i \leq n} \|X - Y_i\| \xrightarrow{D} \nu,$$

where ν in a scale mixture of Weibull distributions with distribution function

$$F_\nu(t) = \begin{cases} 0, & t \leq 0 \\ 1 - \int \exp(-\lambda^d(B(0,1))g(x)t^d)\, dP(x), & t > 0. \end{cases}$$

If additionally $E\|X - Y_1\|^r < \infty$, then $\int g^{-r/d}\, dP < \infty$ and

$$\lim_{n \to \infty} n^{r/d} E \min_{1 \leq i \leq n} \|X - Y_i\|^r = \Gamma\Big(2 + \frac{r}{d}\Big) M_r(B(0,1)) \int g^{-r/d}\, dP$$

if and only if $\left(n^{r/d} \min_{1 \leq i \leq n} \|X - Y_i\|^r\right)_{n \geq 1}$ is uniformly integrable.

(b) *Assume $P(g = 0) > 0$. Then the sequence $(n^{1/d} \min_{1 \leq i \leq n} \|X - Y_i\|)_{n \geq 1}$ is stochastically unbounded.*

Proof

(a) Let $t > 0$. For $x \in \mathbb{R}^d$, we have

$$\mathbb{P}(n^{1/d} \min_{1 \leq i \leq n} \|x - Y_i\| \leq t) = 1 - [1 - Q(B(x, tn^{-1/d}))]^n.$$

By differentation of measures

$$\frac{nQ(B(x, tn^{-1/d}))}{t^d \lambda^d(B(0,1))} = \frac{Q(B(x, tn^{-1/d}))}{\lambda^d(B(x, tn^{-1/d}))} \to g(x)$$

as $n \to \infty$ for λ^d-almost all x and hence for P-almost all $x \in \mathbb{R}^d$ (cf. Chatterji, 1973, Chapitre V). Therefore, by Lebesgue's dominated convergence theorem

$$\mathbb{P}(n^{1/d} \min_{1\le i\le n} \|X - Y_i\| \le t) = \int \mathbb{P}(n^{1/d} \min_{1\le i\le n} \|x - Y_i\| \le t)\, dP(x) \to F_\nu(t).$$

Obviously, $\mathbb{P}(n^{1/d} \min_{1\le i\le n} \|X - Y_i\| = 0) = 0 = F_\nu(0)$. Since $P(g > 0) = 1$, F_ν is a distribution function and we obtain

$$n^{1/d} \min_{1\le i\le n} \|X - Y_i\| \xrightarrow{D} \nu.$$

This, together with

$$\begin{aligned}\int z^r \, d\nu(z) &= \Gamma(1 + \frac{r}{d})(\lambda^d(B(0,1))^{-r/d} \int g^{-r/d}\, dP \\ &= \Gamma(2 + \frac{r}{d}) M_r(B(0,1)) \int g^{-r/d}\, dP\end{aligned}$$

implies the second assertion (cf. Hoffmann-Jørgensen, 1994, 5.2).

(b) As above (but now F_ν is not a distribution function)

$$\mathbb{P}(n^{1/d} \min_{1\le i\le n} \|X - Y_i\| > t) \to 1 - F_\nu(t) \ge P(g = 0) > 0$$

for every $t \ge 0$. □

For $P = h\lambda^d$ such that $h \in L_{d/(d+r)}(\lambda^d)$, let $P_r = h_r\lambda^d$ be defined as in (7.5). Set

$$B = \left\{ g \in L_1(\lambda^d) \colon g \ge 0, \int g \, d\lambda^d \le 1 \right\}.$$

Then the function

$$F \colon B \to [0, \infty], \quad F(g) = \int g^{-r/d}\, dP$$

satisfies

$$F(h_r) = \min_{g \in B} F(g) = \|h\|_{d/(d+r)} \tag{9.1}$$

(cf. Lemma 6.8). In fact, by Hölder's inequality with $p = d/(d + r)$ and $q = -d/r$, one obtains

$$\begin{aligned}F(g) &\ge \|h\|_p \left(\int (g^{-r/d})^q \, d\lambda^d \right)^{1/q} \\ &\ge \|h\|_p = F(h_r)\end{aligned}$$

for $g \in B$. This shows that an asymptotically optimal random quantizer for P of order r is given by the distribution $Q = P_r$ provided the uniform integrability condition holds for P_r. In this case Theorem 9.1 yields

$$\lim_{n\to\infty} n^{r/d} E \min_{1\le i\le n} \|X - Y_i\|^r = \Gamma\left(2 + \frac{r}{d}\right) M_r(B(0,1)) \|h\|_{d/(d+r)}. \tag{9.2}$$

We will not discuss uniform integrability of $\left(n^{r/d} \min_{1\le i\le n} \|X - Y_i\|^r\right)_{n\ge 1}$ in general but consider only the case of uniformly distibuted Y_i.

9.2 Theorem
Let $A \subset \mathbb{R}^d$ be a compact set with $\lambda^d(A) > 0$ and let $Y_1, Y_2, \ldots$ be i.i.d. $U(A)$-distributed random variables independent of X. Assume $P = P_a$, $\operatorname{supp}(P) \subset A$ or $P(\operatorname{int} A) = 1$. Assume further that there are constants $c > 0$ and $t_0 > 0$ such that

$$\lambda^d(B(x,t) \cap A) \geq ct^d \text{ for every } x \in \operatorname{supp}(P),\ t \in (0, t_0). \tag{9.3}$$

Then

$$n^{1/d} \min_{1 \leq i \leq n} \|X - Y_i\| \xrightarrow{D} \nu$$

and

$$\lim_{n \to \infty} n^{r/d} E \min_{1 \leq i \leq n} \|X - Y_i\|^r = \Gamma\Big(2 + \frac{r}{d}\Big) M_r(B(0,1)) \lambda^d(A)^{r/d},$$

where ν denotes the Weibull distribution with distribution function

$$F_\nu(t) = \begin{cases} 0, & t < 0 \\ 1 - \exp(-\lambda^d(B(0,1))t^d / \lambda^d(A)), & t \geq 0. \end{cases}$$

Proof

Let $Q = U(A)$. In case $P = P_a$ and $\operatorname{supp}(P) \subset A$, the first assertion follows from Theorem 9.1. In case $P(\operatorname{int} A) = 1$, we have for $t \geq 0$ and $x \in \operatorname{int} A$

$$\mathbb{P}\,(n^{1/d} \min_{1 \leq i \leq n} \|x - Y_i\| \leq t) = 1 - \left(1 - \frac{\lambda^d(B(0,1))t^d}{n \lambda^d(A)}\right)^n$$

for sufficiently large n. This implies

$$n^{1/d} \min_{1 \leq i \leq n} \|X - Y_i\| \xrightarrow{D} \nu.$$

Furthermore

$$\sup_{n \geq 1} n^{s/d} E \min_{1 \leq i \leq n} \|X - Y_i\|^s < \infty \text{ for every } s \in [1, \infty). \tag{9.4}$$

This property implies that $\left(n^{r/d} \min_{1 \leq i \leq n} \|X - Y_i\|^r\right)_{n \geq 1}$ is uniformly integrable which yields the second assertion.

To prove (9.4) note that $\operatorname{supp}(P) \subset A$. Let $x \in \operatorname{supp}(P)$. We have

$$\begin{aligned} E \min_{1 \leq i \leq n} \|x - Y_i\|^s &= \int_0^\infty \mathbb{P}(\min_{1 \leq i \leq n} \|x - Y_i\| > t^{1/s})\, dt \\ &= s \int_0^{\operatorname{diam}(A)} [1 - Q(B(x,t))]^n t^{s-1}\, dt. \end{aligned}$$

Observe that (9.3) holds (with a different constant c) for every $t \in (0, \operatorname{diam}(A))$ if $\operatorname{diam}(A) > t_0$. In fact, choose $t_1 \in (0, t_0)$. Then for every $t_0 \leq t < \operatorname{diam}(A)$

$$\lambda^d(B(x,t) \cap A) \geq \lambda^d(B(x,t_1) \cap A) \geq ct_1^d \geq c(t_1/\operatorname{diam}(A))^d t^d.$$

Using the inequality $(1-z)^n \leq e^{-nz}$, $0 \leq z \leq 1$, one obtains with $c_1 = c/\lambda^d(A)$

$$\begin{aligned} E \min_{1 \leq i \leq n} \|x - Y_i\|^s &\leq \int_0^{\operatorname{diam}(A)} (1 - c_1 t^d)^n t^{s-1}\, dt \\ &\leq s \int_0^\infty \exp(-nc_1 t^d) t^{s-1}\, dt \\ &= \frac{s\Gamma(s/d)}{d(nc_1)^{s/d}} =: c_2 n^{-s/d}. \end{aligned}$$

This gives

$$\sup_{n \geq 1} n^{s/d} E \min_{1 \leq i \leq n} \|X - Y_i\|^s = \sup_{n \geq 1} n^{s/d} \int E \min_{1 \leq i \leq n} \|x - Y_i\|^s\, dP(x) \leq c_2.$$

□

The regularity condition (9.3) with supp(P) replaced by supp($U(A)$) is discussed in Section 12. Compact convex subsets A of $\mathbb{R}^d$ with $\lambda^d(A) > 0$ satisfy this condition.

Next we will deal with consequences for the quantization coefficients.

9.2 Random quantizer upper bound

Clearly, random quantizers cannot be better than optimal quantizers giving the following upper bound for $Q_r([0,1]^d)$. For the l_2-norm, it is due to Zador (1963, 1982).

9.3 Proposition

$$Q_r([0,1]^d) \leq \Gamma\left(2 + \frac{r}{d}\right) M_r(B(0,1)).$$

Proof

Choose $A = [0,1]^d$ and $P = U([0,1]^d)$ in Theorem 9.2. Then

$$M_{n,r}([0,1]^d) \leq E \min_{1 \leq i \leq n} \|X - Y_i\|^r,\ n \geq 1$$

which yields the assertion. □

A comparison of the random quantizer and the cube quantizer upper bound shows for the l_2-norm and $r = 2$

$$\Gamma\Big(2+\frac{2}{d}\Big) M_2(B(0,1)) < M_2([0,1]^d), \; d \geq 7.$$

For the l_1-norm and $r = 1$ one obtains

$$\Gamma\Big(2+\frac{1}{d}\Big) M_1(B(0,1)) < M_1([0,1]^d), \; d \geq 5.$$

A comparison of the random quantizer upper bound and the ball lower bound given in Proposition 8.3 can be found in the Tables 9.1 and 9.2.

d	$M_2(B(0,1))$	$\Gamma(2+\frac{2}{d})M_2(B(0,1))$
1	0.0833	0.5
2	0.1592	0.3183
3	0.2309	0.3471
4	0.3001	0.3989
5	0.3676	0.4566
6	0.4338	0.5165
7	0.4991	0.5774
8	0.5636	0.6386
9	0.6276	0.7000
10	0.6910	0.7614
20	1.3105	1.3714
30	1.9169	1.9744
40	2.5175	2.5733
50	3.1149	3.1696
100	6.0801	6.1325

Table 9.1: l_2-norm, $r = 2$. Ball lower bound and random quantizer upper bound for $Q_2([0,1]^d)$

d	$M_1(B(0,1))$	$\Gamma(2+\frac{1}{d})M_1(B(0,1))$
1	0.25	0.5
2	0.4714	0.6267
3	0.6814	0.8113
4	0.8853	1.0031
5	1.0855	1.1960
6	1.2831	1.3887
7	1.4788	1.5809
8	1.6730	1.7725
9	1.8662	1.9636
10	2.0585	2.1542
20	3.9545	4.0422
30	5.8280	5.9128
40	7.6920	7.7526
50	9.5506	9.6330
100	18.8083	18.8886

Table 9.2: l_1-norm, $r = 1$

9.3 d-asymptotics and entropy

From the ball lower bound and the random quantizer upper bound for $Q_r([0,1]^d)$ we deduce the following approximation for large d.

9.4 Corollary
Let the underlying norm be the l_p-norm, $1 \le p < \infty$. Then

$$\lim_{d\to\infty} d^{-r/p} Q_r([0,1]^d) = \frac{p^r}{2^r (ep)^{r/p}\, \Gamma(\frac{1}{p})^r}.$$

Proof

By (2.5) and (2.6)

$$M_r(B(0,1)) = \frac{d\Gamma(1+\frac{d}{p})^{r/d}}{(d+r)2^r\Gamma(1+\frac{1}{p})^r}.$$

From Stirling's formula for the Γ-function, i.e.

$$\Gamma(x) \sim \sqrt{2\pi} x^{x-\frac{1}{2}} e^{-x} \text{ as } x \to \infty,$$

we deduce

$$\lim_{d\to\infty} d^{-r/p} \Gamma\Big(1+\frac{d}{p}\Big)^{r/d} = (ep)^{-r/p}.$$

Since $\lim_{d\to\infty}\Gamma(2+\frac{r}{d}) = \Gamma(2) = 1$, the assertion follows from the Propositions 8.3 and 9.3. □

For the l_2-norm and $r = 2$, one obtains from the preceding corollary

$$\lim_{d\to\infty} d^{-1}Q_2([0,1]^d) = \frac{1}{2\pi e} = 0.0585\ldots.$$

In this case the cube quantizer yields

$$d^{-1}M_2([0,1]^d) = \frac{1}{12} = 0.0833\ldots,\ d \geq 1$$

and the lattice quantizer based on the lattice D_d gives

$$\lim_{d\to\infty} d^{-1}M_2(W(0|D_d)) = \frac{1}{12}$$

(cf. Example 8.13). For the l_1-norm and $r = 1$, one gets

$$\lim_{d\to\infty} d^{-1}Q_1([0,1]^d) = \frac{1}{2e} = 0.1839\ldots,$$
$$d^{-1}M_1([0,1]^d) = \frac{1}{4},\ d \geq 1,$$
$$\lim_{d\to\infty} d^{-1}M_1(W(0|D_d)) = \frac{1}{4}.$$

The d-asymptotics for the quantization coefficients of arbitrary product measures with identical one-dimensional marginals now follows from the preceding corollary and the well known fact that the Renyi entropy of order s approaches the Shannon differential entropy as $s \to 1$. For $P = h\lambda^d$, the **Renyi entropy of order s** is defined by

$$H_s(P) = \frac{1}{1-s}\log\int h^s\,d\lambda^d,\ 0 < s < 1. \tag{9.5}$$

The **differential entropy** is defined by

$$H(P) = -\int h\log h\,d\lambda^d = -\int \log h\,dP \tag{9.6}$$

provided the integral exists. Note that in (9.5) and (9.6) the entropies are calculated in nats and not in bits.

For the l_2-norm, the following result is contained in Zador (1963). □

9.5 Proposition
Let $X_1, X_2, \ldots$ be i.i.d. real random variables with distribution P. Suppose $P_a \neq 0$ and $E|X_1|^{r+\delta} < \infty$ for some $\delta > 0$. Let the underlying norm be the l_p-norm, $1 \leq p < \infty$. Then

$$\lim_{d\to\infty} Q_r(X_1,\ldots,X_d) = 0 \text{ if } P_a \neq P,$$

$$\lim_{d\to\infty} d^{-r/p} Q_r(X_1,\dots,X_d) = \frac{p^r e^{rH(P)}}{2^r (ep)^{r/p}\Gamma(\frac{1}{p})^r} \text{ if } P_a = P.$$

Proof

It follows from the assumptions that

$$\left(\bigotimes_1^d P\right)_a = \bigotimes_1^d P_a \neq 0 \text{ and } E\|(X_1,\dots,X_d)\|^{r+\delta} < \infty.$$

The r-th quantization coefficient of $(X_1,\dots,X_d)$ is given by

$$Q_r(X_1,\dots,X_d) = Q_r([0,1)^d) P_a(\mathbb{R})^d \left\|\bigotimes_1^d h\right\|_{d/(d+r)}$$

where $h = d\tilde{P}_a/d\lambda^d$ and $\tilde{P}_a = P_a/P_a(\mathbb{R})$. Since $h \in L_{1/(1+r)}(\lambda)$ by Remark 6.3, the differential entropy $H(\tilde{P}_a)$ is well defined and $H(\tilde{P}_a) \in [-\infty,\infty)$ (cf. Vajda, 1989, p. 316). We have

$$\begin{aligned}\log\left\|\bigotimes_1^d h\right\|_{d/(d+r)} &= d\log\|h\|_{d/(d+r)} \\ &= (d+r)\log\int h^{d/(d+r)}\,d\lambda \\ &= rH_{d/(d+r)}(\tilde{P}_a).\end{aligned}$$

Therefore

$$\lim_{d\to\infty}\log\left\|\bigotimes_1^d h\right\|_{d/(d+r)} = rH(\tilde{P}_a).$$

Furthermore

$$\lim_{d\to\infty} d^{r/p} P_a(\mathbb{R})^d = 0 \text{ if } P_a(\mathbb{R}) < 1.$$

Thus both assertions folow from Corollary 9.4. □

We see that $d^{r/p}$ is the correct order of convergence of $Q_r(X_1,\dots,X_d)$ to infinity (under the l_p-norm) provided $P = P_a$ and $H(P) \in \mathbb{R}$.

9.6 Remark

(a) Proposition 9.5 immediately yields the d-asymptotics for the vector quantizer advantage as defined in (7.8) in the i.i.d. case:

$$\text{(9.7)} \qquad \lim_{d\to\infty} \frac{dQ_r(X_1)}{Q_r(X_1,\dots,X_d)} = \frac{2^r er\Gamma(\frac{1}{r})^r Q_r(X_1)}{r^r e^{rH(X_1)}} \geq 1.$$

(Here the underlying norm in the l_r-norm, $E|X_1|^{r+\delta} < \infty$ for some $\delta > 0$, and the distribution of X_1 is absolutely continuous with respect to λ.)

(b) We know from Table 8.2 that $\inf Q_2(P) = 0$ and $\sup Q_2(P) = \infty$, where the infimum and the supremum are taken over all univariate symmetric (absolutely continuous) distributions P with variance equal to one. On the other hand, the normal distribution $N(0,1)$ is the unique maximizer of the differential entropy among all probabilities P with mean zero, variance equal to one and $\text{supp}(P) = \mathbb{R}$. For such distributions P it follows from Proposition 9.5 that

$$Q_2\left(\bigotimes_1^d P\right) < Q_2\left(\bigotimes_1^d N(0,1)\right) \tag{9.8}$$

for sufficiently large d. Consider for instance, the hyper-exponential distribution $P = HE(a,b)$ with $a^2 = \Gamma(\frac{1}{b})/\Gamma(\frac{3}{b})$ and $b = 1/10$. Then $Q_2(P) = 37.0908\ldots$ while $Q_2(N(0,1)) = 2.7206\ldots$ (cf. Table 8.2). The inequality (9.8) holds for $d \geq 2$ (cf. Table 9.4).

Anlogous statements are valid for distributions which maximize the differential entropy among other classes of univariate distributions.

Notes

A discussion of the uniform integrability condition in Theorem 9.1 can be found in Zador (1963) and Stadje (1995). Gersho (1979) contains an extension of Proposition 9.5 for the l_2-norm to stationary ergodic sequences.

P	$H(P)$
$N(0,\sigma^2)$	$\frac{1}{2}\log(2\pi\sigma^2)+\frac{1}{2}$
$L(a)$	$\log a + 2$
$DE(a)$	$\log(2a)+1$
$D\Gamma(a,b)$	$\log(2a\Gamma(b))+(1-b)\psi(b)+b$
$HE(a,b)$	$\log(\frac{2a\Gamma(1/b)}{b})+\frac{1}{b}$
$U([a,b])$	$\log(b-a)$
$E(a)$	$\log a + 1$
$\Gamma(a,b)$	$\log(a\Gamma(b))+(1-b)\psi(b)+b$
$W(a,b)$	$\log(\frac{a}{b})+\frac{(b-1)\gamma}{b}+1$
$P(a,b)$	$\log(\frac{a}{b})+\frac{b+1}{b}$

Table 9.3: Differential entropies. $\psi=\Gamma'/\Gamma$, $\gamma=$ Euler's constant $=0.5772\ldots$

P	$\frac{Q_r(\bigotimes^{d} P)}{Q_r([0,1]^d)}=\exp(rH_{d/(d+r)}(P))$
$N(0,\sigma^2)$	$(2\pi\sigma^2)^{r/2}(\frac{d+r}{d})^{(d+r)/2}$
$L(a)$	$a^r(\frac{\Gamma(\frac{d}{d+r})^2}{\Gamma(\frac{2d}{d+r})})^{d+r}$
$DE(a)$	$(2a)^r(\frac{d+r}{d})^{d+r}$
$D\Gamma(a,b)$	$\frac{(2a)^r}{\Gamma(b)^d}(\frac{d+r}{d})^{bd+r}\Gamma(\frac{bd+r}{d+r})^{d+r}$
$HE(a,b)$	$(\frac{2a\Gamma(\frac{1}{b})}{b})^r(\frac{d+r}{d})^{(d+r)/b}$

Table 9.4: Quantization coefficients for product probability measures up to $Q_r([0,1]^d)$

10 Asymptotics for the covering radius

Let P be a Borel probability measure on $\mathbb{R}^d$ and let $\| \ \|$ denote any norm on $\mathbb{R}^d$. For $1 \le r \le \infty$ and $g : \mathbb{R}^d \to \mathbb{R}$ Borel measurable, define

$$\|g\|_{P,r} = \|g\|_r = \begin{cases} \left(\int |g|^r \, dP\right)^{1/r}, & 1 \le r < \infty \\ \inf\{c \ge 0 : |g| \le c \ P\text{-a.s.}\}, & r = \infty \end{cases}$$

and

(10.1)
$$e_{n,r}(P) = \inf_{\substack{\alpha \subset \mathbb{R}^d \\ |\alpha| \le n}} \|d_\alpha\|_r,$$

where $d_\alpha(x) = d(x, \alpha) = \inf_{a \in \alpha} \|x - a\|$. It follows from Lemma 3.1 that, for $1 \le r < \infty$,

(10.2)
$$V_{n,r}(P)^{1/r} = e_{n,r}(P).$$

In this section we consider the case $r = \infty$ and discuss its relation to the quantization problem ($r < \infty$).

10.1 Basic properties

Note that $e_{n,\infty}(P) < \infty$ if $\mathrm{supp}(P)$ is compact. It follows from the continuity of d_α that $\|d_\alpha\|_\infty = \sup_{x \in \mathrm{supp}(P)} d_\alpha(x)$ and hence

$$e_{n,\infty}(P) = \inf_{\substack{\alpha \subset \mathbb{R}^d \\ |\alpha| \le n}} \sup_{x \in \mathrm{supp}(P)} \min_{a \in \alpha} \|x - a\|.$$

So, if we define for a nonempty compact set $A \subset \mathbb{R}^d$,

(10.3)
$$e_{n,\infty}(A) = \inf_{\substack{\alpha \subset \mathbb{R}^d \\ |\alpha| \le n}} \max_{x \in A} \min_{a \in \alpha} \|x - a\|$$

then $e_{n,\infty}(P) = e_{n,\infty}(A)$ for every probability P with $\mathrm{supp}(P) = A$. A set $\alpha \subset \mathbb{R}^d$ with $|\alpha| \le n$ for which the above infimum is attained, is called an **n-optimal set of centers for A of order ∞**. Let $C_{n,\infty}(A)$ denote the set of all n-optimal sets of centers for A of order ∞. Since

$$\max_{x \in A} \min_{a \in \alpha} \|x - a\| = \min\Big\{s \ge 0 : \bigcup_{a \in \alpha} B(a, s) \supset A\Big\},$$

searching for $\alpha \in C_{n,\infty}(A)$ is equivalent to the geometric problem of finding the most economical covering of A by at most n balls of equal radius. The number $e_{n,\infty}(A)$ is called **n-th covering radius for A**. Recall that the Hausdorff metric is given by

$$d_H(A, B) = \max\{\max_{a \in A} \min_{b \in B} \|a - b\|, \max_{b \in B} \min_{a \in A} \|a - b\|\}$$

for nonempty compact sets $A, B \subset \mathbb{R}^d$.

10.1 Lemma
Let $n \in \mathbb{N}$.

(a) *If $1 \leq r \leq s \leq \infty$, then $e_{n,r}(P) \leq e_{n,s}(P)$.*

(b) *If $\operatorname{supp}(P)$ is compact, then $\lim_{r \to \infty} e_{n,r}(P) = e_{n,\infty}(P)$.*

(c) *Let A denote the support of P and suppose A is compact and $|A| \geq n$. Let $\alpha_r \in C_{n,r}(P)$, $1 \leq r < \infty$, and let $(r_k)_{k \geq 1}$ be a sequence in $[1, \infty)$ converging to infinity. Then the set of d_H-cluster points of the sequence $(\alpha_{r_k})_{k \geq 1}$ is a nonempty subset of $C_{n,\infty}(A)$ and*

$$\lim_{r \to \infty} d_H(\alpha_r, C_{n,\infty}(A)) = 0.$$

Proof

(a) follows from the fact that $\| \ \|_r \leq \| \ \|_s$ for $r \leq s$.
(b) and (c). Let $u = \operatorname{diam}(A)$ and choose $s > 0$ such that $A \subset B(0, s)$. It follows from Theorem 4.1 and Lemma 2.6 that

$$C_{n,r}(P) \subset \{\alpha \subset \mathbb{R}^d : 1 \leq |\alpha| \leq n, \ \alpha \subset B(0, s + u)\}$$

for every $r \in [1, \infty)$ provided $|A| \geq n$. Using Lemma 4.23 we deduce the existence of a d_H-cluster point for $(\alpha_{r_k})_{k \geq 1}$. Now let α be the d_H-limit of a subsequence of $(\alpha_{r_k})_{k \geq 1}$ which is again denoted by $(\alpha_{r_k})_{k \geq 1}$. We have

$$\begin{aligned} \|d_{\alpha_{r_k}} - d_\alpha\|_{r_k} &\leq \|d_{\alpha_{r_k}} - d_\alpha\|_\infty \\ &\leq \sup_{x \in \mathbb{R}^d} |d(x, \alpha_{r_k}) - d(x, \alpha)| \\ &= d_H(\alpha_{r_k}, \alpha) \end{aligned}$$

and hence

$$\begin{aligned} \|d_{\alpha_{r_k}}\|_{r_k} &\geq \|d_\alpha\|_{r_k} - \|d_{\alpha_{r_k}} - d_\alpha\|_{r_k} \\ &\geq \|d_\alpha\|_{r_k} - d_H(\alpha_{r_k}, \alpha). \end{aligned}$$

By (a), $\lim_{r \to \infty} e_{n,r}(P)$ exists and is less than or equal to $e_{n,\infty}(P)$. Therefore

$$\begin{aligned} e_{n,\infty}(P) &\geq \lim_{k \to \infty} e_{n,r_k}(P) = \lim_{k \to \infty} \|d_{\alpha_{r_k}}\|_{r_k} \\ &\geq \lim_{k \to \infty} \|d_\alpha\|_{r_k} = \|d_\alpha\|_\infty \geq e_{n,\infty}(P). \end{aligned}$$

This gives (b) and the first assertion of (c). The second assertion of (c) follows from the first one. □

For $A \subset \mathbb{R}^d$ compact with $\lambda^d(A) > 0$, set

$$M_{n,\infty}(A) = \frac{e_{n,\infty}(A)}{\lambda^d(A)^{1/d}}. \tag{10.4}$$

In spite of the slight inconsistency in notation we continue to write $M_\infty(A)$, $M_{n,\infty}(A)$, $Q_\infty(P)$ etc. for the corresponding notions in case $r = \infty$.

10.2 Lemma
Let $A \subset \mathbb{R}^d$ be a nonempty compact set and let $T: \mathbb{R}^d \to \mathbb{R}^d$ be a similarity transformation with scaling number $c > 0$.

(a) $C_{n,\infty}(T(A)) = TC_{n,\infty}(A)$.

(b) $e_{n,\infty}(T(A)) = ce_{n,\infty}(A)$.

(c) $M_{n,\infty}(T(A)) = M_{n,\infty}(A)$ *if* $\lambda^d(A) > 0$.

Proof

Obvious. □

The existence of n-optimal sets of centers of order ∞ can be derived from the existence of n-optimal sets of centers of order $r < \infty$ (cf. Theorem 4.12) and Lemma 10.1(c).

10.3 Lemma (Existence)
If $A \subset \mathbb{R}^d$ is a nonempty compact set, then

$$C_{n,\infty}(A) \neq \emptyset.$$

Proof

We assume without loss of generality that $|A| \geq n$. To show that the assertion follows from Lemma 10.1(c) it suffices to note that $A = \text{supp}(P)$ for some Borel probability measure P on $\mathbb{R}^d$. If A is finite, set $P = \sum_{a \in A} \delta_a / |A|$. Otherwise let $B = \{b_1, b_2, \dots\}$ be a countable dense subset of A and set $P = \sum_{n=1}^{\infty} 2^{-n} \delta_{b_n}$. Then $A = \text{supp}(P)$. □

The covering problem can be formulated in terms of the Hausdorff metric and the L_∞-minimal metric.

10.4 Lemma
Let $A \subset \mathbb{R}^d$ be a nonempty compact set. Then

$$e_{n,\infty}(A) = \inf_{|\alpha| \leq n} d_H(\alpha, A).$$

If $e_{n,\infty}(A) < e_{n-1,\infty}(A)$ ($e_{0,\infty}(A) := \infty$), then

$$C_{n,\infty}(A) = \{\alpha \subset \mathbb{R}^d : 1 \leq |\alpha| \leq n,\ d_H(\alpha, A) = e_{n,\infty}(A)\}.$$

Proof

Let $\alpha \in C_{n,\infty}(A)$ and set

$$\beta = \{a \in \alpha : d(a, A) \leq e_{n,\infty}(A)\}.$$

Then $\beta \neq \emptyset$ and

$$e_{n,\infty}(A) = \max_{x \in A} d(x, \alpha) = \max_{x \in A} d(x, \beta) = d_H(\beta, A).$$

This yields

$$e_{n,\infty}(A) \geq \inf_{|\alpha| \leq n} d_H(\alpha, A).$$

The converse inequality is obvious. Furthermore, the inclusion

$$C_{n,\infty}(A) \supset \tilde{C} := \{\alpha \subset \mathbb{R}^d : 1 \leq |\alpha| \leq n, d_H(\alpha, A) = e_{n,\infty}(A)\}$$

is also obvious and $\beta \in \tilde{C}$. Now assume $e_{n,\infty}(A) < e_{n-1,\infty}(A)$. Since $\beta \in C_{n,\infty}(A)$, one gets $|\beta| = n$. This gives $\alpha = \beta$ and thus $\alpha \in \tilde{C}$. □

The $\mathbf{L}_\infty$**-minimal metric** ρ_∞ is given by

$$\rho_\infty(P_1, P_2) = \inf\{\varepsilon > 0 : P_1(B) \leq P_2(d_B \leq \varepsilon) \text{ for all } B \in \mathcal{B}(\mathbb{R}^d)\}$$

for Borel probability measures P_1, P_2 on $\mathbb{R}^d$ with compact support.

10.5 Lemma

Suppose supp(P) *is compact and let* X *be a* $\mathbb{R}^d$*-valued random variable with distribution* P. *Then*

$$\begin{aligned} e_{n,\infty}(P) &= \inf_{Q \in \mathcal{P}_n} \rho_\infty(P, Q) \\ &= \inf_{f \in \mathcal{F}_n} \rho_\infty(P, P^f) \\ &= \inf_{f \in \mathcal{F}_n} \operatorname{ess\,sup} \|X - f(X)\|. \end{aligned}$$

Proof

Let A denote the support of P. If $Q \in \mathcal{P}_n$ with $Q(\alpha) = 1$, $|\alpha| \leq n$, let $\varepsilon > 0$ such that $Q(B) \leq P(d_B \leq \varepsilon)$ for all $B \in \mathcal{B}(\mathbb{R}^d)$. Then

$$1 = Q(\alpha) = P(d_\alpha \leq \varepsilon)$$

which gives $A \subset \{d_\alpha \leq \varepsilon\}$. Therefore

$$\|d_\alpha\| = \max_{x \in A} d(x, \alpha) \leq \rho_\infty(P, Q).$$

This implies

$$e_{n,\infty}(P) \leq \inf_{Q \in \mathcal{P}_n} \rho_\infty(P, Q) \leq \inf_{f \in \mathcal{F}_n} \rho_\infty(P, P^f).$$

If $\alpha \subset \mathbb{R}^d$ with $1 \leq |\alpha| \leq n$, let $\varepsilon = \max_{x \in A} d(x, \alpha)$. Choose a Voronoi partition $\{A_a : a \in \alpha\}$ of $\mathbb{R}^d$ with respect to α and let $f = \sum_{a \in \alpha} a 1_{A_a}$. Since $A_a \cap A \subset B(a, \varepsilon)$ for every $a \in \alpha$, one obtains for $\beta \subset \alpha$

$$P^f(\beta) = P\left(\bigcup_{a \in \beta} A_a\right) \leq P\left(\bigcup_{a \in \beta} B(a, \varepsilon)\right) = P(d_\beta \leq \varepsilon).$$

Therefore

$$\rho_\infty(P, P^f) \le \varepsilon = \max_{x \in A} d(x, \alpha).$$

This implies

$$\inf_{f \in \mathcal{F}_n} \rho_\infty(P, P^f) \le e_{n,\infty}(P).$$

For $f \in \mathcal{F}_n$, let $\alpha = f(\mathbb{R}^d)$ and $A_a = \{f = a\}$, $a \in \alpha$. Then

$$\begin{aligned}
&\operatorname{ess\,sup} \|X - f(X)\| \\
&= \inf\Big\{c \ge 0 \colon \sum_{a \in \alpha} P\big(A_a \cap \{x \in \mathbb{R}^d \colon \|x - a\| > c\}\big) = 0\Big\} \\
&\ge \inf\Big\{c \ge 0 \colon \sum_{a \in \alpha} P\big(A_a \cap \{d_\alpha > c\}\big) = 0\Big\} \\
&= \|d_\alpha\|_\infty,
\end{aligned}$$

and if $\{A_a \colon a \in \alpha\}$ is a Voronoi partition of $\mathbb{R}^d$ with respect to α, then

$$\operatorname{ess\,sup} \|X - f(X)\| = \|d_\alpha\|_\infty.$$

This implies

$$e_{n,\infty}(P) = \inf_{f \in \mathcal{F}_n} \operatorname{ess\,sup} \|X - f(X)\|.$$

□

Since $\rho_\infty(P_1, P_2) \ge d_H(\operatorname{supp}(P_1), \operatorname{supp}(P_2))$, ρ_∞-convergence implies weak convergence and d_H-convergence of the supports.

10.2 Asymptotic covering radius

Clearly, if $A \subset \mathbb{R}^d$ is nonempty compact then $e_{n,\infty}(A)$ decreases to zero as $n \to \infty$. We need the following simple lemma .

10.6 Lemma

(a) If $A, B \subset \mathbb{R}^d$ are nonempty compact sets with $A \subset B$, then $e_{n,\infty}(A) \le e_{n,\infty}(B)$.

(b) If $A_i \subset \mathbb{R}^d$ are nonempty compact sets and $n_i \in \mathbb{N}$ with $\sum_{i=1}^m n_i \le n$, then $e_{n,\infty}(\bigcup_{i \le 1}^m A_i) \le \max_{1 \le i \le m} e_{n_i,\infty}(A_i)$.

Proof

(a) Let $\beta \in C_{n,\infty}(B)$. Then

$$e_{n,\infty}(A) \le \max_{x \in A} \min_{b \in \beta} \|x - b\| \le e_{n,\infty}(B).$$

(b) Let $\alpha_i \in C_{n_i,\infty}(A_i)$ and let $\alpha = \bigcup_{i=1}^{m} \alpha_i$. Then $|\alpha| \leq n$ and therefore,

$$\begin{aligned} e_{n,\infty}(\bigcup_{i=1}^{m} A_i) &\leq \max_{x \in \bigcup_{i=1}^{m} A_i} \min_{a \in \alpha} \|x - a\| \\ &= \max_{1 \leq i \leq m} \max_{x \in A_i} \min_{a \in \alpha} \|x - a\| \\ &\leq \max_{1 \leq i \leq m} \max_{x \in A_i} \min_{a \in \alpha_i} \|x - a\| \\ &= \max_{1 \leq i \leq m} e_{n_i,\infty}(A_i). \end{aligned}$$

□

Now we can derive the exact asymptotic first order behaviour of the covering radius $e_{n,\infty}(A)$ for compact Jordan measurable sets A with $\lambda^d(A) > 0$.

10.7 Theorem (Asymptotic covering radius)
Let $A \subset \mathbb{R}^d$ be a nonempty compact set with $\lambda^d(\partial A) = 0$. Let

$$Q_\infty([0,1]^d) = \inf_{n \geq 1} n^{1/d} e_{n,\infty}([0,1]^d).$$

Then $Q_\infty([0,1]^d) > 0$ and

$$\lim_{n \to \infty} n^{1/d} e_{n,\infty}(A) = Q_\infty([0,1]^d) \lambda^d(A)^{1/d}.$$

Proof

The proof is given in three steps.

Step 1. Let $A = [0,1]^d$. Let $m, n \in \mathbb{N}$, $m < n$ and let $k = k(n,m) = [(\frac{n}{m})^{1/d}]$. Choose a tesselation of the unit cube $[0,1]^d$ consisting of k^d translates $C_1, \ldots, C_{k^d}$ of the cube $[0, \frac{1}{k}]^d$. Then by Lemmas 10.2 and 10.6,

$$\begin{aligned} e_{n,\infty}([0,1]^d) &\leq \max_{1 \leq i \leq k^d} e_{m,\infty}(C_i) \\ &= \max_{1 \leq i \leq k^d} M_{m,\infty}(C_i) k^{-1} \\ &= k^{-1} M_{m,\infty}([0,1]^d) \\ &= k^{-1} e_{m,\infty}([0,1]^d) \end{aligned}$$

and hence

$$n^{1/d} e_{n,\infty}([0,1]^d) \leq \frac{k+1}{k} m^{1/d} e_{m,\infty}([0,1]^d).$$

This implies

$$\limsup_{n \to \infty} n^{1/d} e_{n,\infty}([0,1]^d) \leq m^{1/d} e_{m,\infty}([0,1]^d)$$

for every $m \in \mathbb{N}$. Therefore, $\lim_{n\to\infty} n^{1/d} e_{n,\infty}([0,1]^d)$ exists in $[0,\infty)$ and

$$\lim_{n\to\infty} n^{1/d} e_{n,\infty}([0,1]^d) = Q_\infty([0,1]^d). \tag{10.5}$$

From the subsequent Proposition 10.10 (a) it follows that $Q_\infty([0,1]^d) > 0$.

Step 2. Let $A = \bigcup_{i=1}^m C_i$, where $\{C_1,\dots,C_m\}$ is a packing in $\mathbb{R}^d$ consisting of closed cubes whose edges are parallel to the coordinate axes and with common length of the edges $l(C_i) = l > 0$. Let $n_1 = n_1(n) = [\frac{n}{m}]$. Then by Lemmas 10.2 and 10.6

$$\begin{aligned} e_{n,\infty}(A) &\le \max_{1\le i\le m} e_{n_1,\infty}(C_i) \\ &= \max_{1\le i\le m} M_{n_1,\infty}(C_i) l \\ &= e_{n_1,\infty}([0,1]^d) l, \; n \ge m. \end{aligned}$$

From Step 1 it follows that

$$n^{1/d} e_{n_1,\infty}([0,1]^d) = \left(\frac{n}{n_1}\right)^{1/d} n_1^{1/d} e_{n_1,\infty}([0,1]^d) \to m^{1/d} Q_\infty([0,1]^d) \text{ as } n \to \infty.$$

This implies

$$\limsup_{n\to\infty} n^{1/d} e_{n,\infty}(A) \le Q_\infty([0,1]^d) m^{1/d} l = Q_\infty([0,1]^d) \lambda^d(A)^{1/d}. \tag{10.6}$$

To prove that $Q_\infty([0,1]^d)\lambda^d(A)^{1/d}$ is a lower bound for $L = \liminf_{n\to\infty} n^{1/d} e_{n,\infty}(A)$, let $\beta = \beta(n) \in C_{n,\infty}(A)$ for $n \in \mathbb{N}$, $\beta_i = \beta_i(n) = \beta \cap \operatorname{int} C_i$ and $n_i = n_i(n) = |\beta_i|$, $1 \le i \le m$. For $0 < \varepsilon < l/2$, let $C_{i,\varepsilon} \subset C_i$ be a parallel closed cube with the same midpoint as C_i and edge-length $l(C_{i,\varepsilon}) = l - 2\varepsilon$. Choose a finite set $\gamma_i = \gamma_i(\varepsilon) \subset C_{i,\varepsilon}$, $|\gamma_i| = k = k(\varepsilon)$ say, such that

$$\min_{a\in\gamma_i} \|x - a\| \le \inf_{y\in C_i^c} \|x - y\| \text{ for every } x \in C_{i,\varepsilon},\ 1 \le i \le m.$$

Then

$$\begin{aligned} e_{n,\infty}(A) &= \max_{x\in A} \min_{b\in\beta} \|x - b\| \\ &\ge \max_{1\le i\le m} \max_{x\in C_i} \min_{b\in\beta\cup\gamma_i} \|x - b\| \\ &\ge \max_{1\le i\le m} \max_{x\in C_{i,\varepsilon}} \min_{b\in\beta\cup\gamma_i} \|x - b\| \\ &= \max_{1\le i\le m} \max_{x\in C_{i,\varepsilon}} \min_{b\in\beta_i\cup\gamma_i} \|x - b\| \\ &\ge \max_{1\le i\le m} e_{n_i+k,\infty}(C_{i,\varepsilon}) \\ &= \max_{1\le i\le m} e_{n_i+k,\infty}([0,1]^d)(l - 2\varepsilon). \end{aligned}$$

Choose a subsequence (also denoted by (n)) such that

$$\frac{n_i}{n} \to v_i \in [0,1],\ 1 \le i \le m \text{ and } n^{1/d} e_{n,\infty}(A) \to L \text{ as } n \to \infty.$$

Since $\sum_{i=1}^{m} n_i \le n$, we have $\sum_{i=1}^{m} v_i \le 1$. Furthermore, $v_i > 0$ for every i. Otherwise, Step 1 yields $L = \infty$, which contradicts (10.6). By taking a further subsequence we can assume without loss of generality that

$$\lim_{n\to\infty} (n_i + k)^{1/d} e_{n_i+k,\infty}([0,1]^d) = Q_\infty([0,1]^d).$$

This implies

$$L \ge \max_{1\le i\le m} Q_\infty([0,1]^d) v_i^{-1/d}(l - 2\varepsilon).$$

Since $0 < \varepsilon < l/2$ is arbitrary and $\max_{1\le i\le m} v_i^{-1/d} \ge m^{1/d}$, one obtains

$$\begin{aligned} L &\ge \max_{1\le i\le m} Q_\infty([0,1]^d) v_i^{-1/d} l \\ &\ge Q_\infty([0,1]^d) m^{1/d} l \\ &= Q_\infty([0,1]^d) \lambda^d(A)^{1/d}. \end{aligned} \tag{10.7}$$

Step 3. Let A be an arbitrary compact subset of $\mathbb{R}^d$. Let $A \subset C$ for some closed cube C whose edges are parallel to the coordinate axes with edge-length $l(C) = l$. For $k \in \mathbb{N}$ consider a tesselation of C consisting of closed cubes $C_1, \ldots, C_{k^d}$ of common edge-length $l(C_i) = l/k$. Set

$$A_k = \bigcup \{C_i : A \cap C_i \neq \emptyset,\ i \le k^d\}$$

Since $A \subset A_k$, it follows from Step 2 and Lemma 10.6 (a) that

$$\begin{aligned} \limsup_{n\to\infty} n^{1/d} e_{n,\infty}(A) &\le \lim_{n\to\infty} n^{1/d} e_{n,\infty}(A_k) \\ &= Q_\infty([0,1]^d) \lambda^d(A_k)^{1/d},\ k \in \mathbb{N}. \end{aligned}$$

If $U \subset \mathbb{R}^d$ is an open set with $A \subset U$, then $A_k \subset U$ for sufficiency large k. Therefore

$$\begin{aligned} \lambda^d(A) &\le \inf_{k\ge1} \lambda^d(A_k) \\ &\le \inf\{\lambda^d(U) : U \subset \mathbb{R}^d \text{ open}, A \subset U\} = \lambda^d(A) \end{aligned}$$

which yields

$$\limsup_{n\to\infty} n^{1/d} e_{n,\infty}(A) \le Q_\infty([0,1]^d) \lambda^d(A)^{1/d}. \tag{10.8}$$

Now suppose $\lambda^d(\partial A) = 0$. To prove that $Q_\infty([0,1]^d)\lambda^d(A)^{1/d}$ is a lower bound for $\liminf_{n\to\infty} n^{1/d} e_{n,\infty}(A)$ we may assume that $\lambda^d(A) > 0$. Set

$$B_k = \bigcup \{C_i : C_i \subset A,\ i \le k^d\}$$

Since $B_k \subset A$, it follows from Step 2 and Lemma 10.6 (a) that

$$\begin{aligned}\liminf_{n\to\infty} n^{1/d} e_{n,\infty}(A) &\geq \lim_{n\to\infty} n^{1/d} e_{n,\infty}(B_k)\\ &= Q_\infty([0,1]^d)\lambda^d(B_k)^{1/d}, \ k \in \mathbb{N} \text{ with } B_k \neq \emptyset.\end{aligned}$$

Since $\lambda^d(\partial A) = 0$, A is Jordan measurable and hence $\sup_{k\geq 1} \lambda^d(B_k) = \lambda^d(A)$. Thus we obtain

$$\liminf_{n\to\infty} n^{1/d} e_{n,\infty}(A) \geq Q_\infty([0,1]^d)\lambda^d(A)^{1/d}. \tag{10.9}$$

Combining (10.8) and (10.9) the theorem is proved. □

For compact sets A with $\lambda^d(A) = 0$ the preceding theorem only yields $e_{n,\infty}(A) = o(n^{-1/d})$. An investigation of the exact order of $e_{n,\infty}(A)$ for several classes of compact sets A with $\lambda^d(A) = 0$ is contained in Chapter III.

For $A \subset \mathbb{R}^d$ compact with $\lambda^d(A) > 0$ and $\lambda^d(\partial A) = 0$, define the **covering coefficient** (with respect to coverings of A by balls of equal radius) by

$$Q_\infty(A) = Q_\infty([0,1]^d)\lambda^d(A)^{1/d}. \tag{10.10}$$

$Q_\infty(A)$ is sometimes called **quantization coefficient of order ∞**. It follows from Lemma 10.1(a), Theorem 6.2 and Theorem 10.5 that $Q_r(P)^{1/r}$ is increasing in r and

$$\lim_{r\to\infty} Q_r(P)^{1/r} = \lim_{r\to\infty} Q_r(A)^{1/r} \leq Q_\infty(A) \tag{10.11}$$

provided $\text{supp}(P) = A$ and P is absolutely continuous with respect to λ^d.

10.8 Remark

(a) We conjecture that equality holds in (10.11) (cf. the special cases given in (10.17), (10.19) and (10.20)). This would show in a precise manner that the covering problem is a limiting case of the quantization problem.

(b) Possibly the condition $\lambda^d(\partial A) = 0$ can be dropped in Theorem 10.7. This is true if the conjecture in (a) can be resolved for the unit cube. In fact, we have for an arbitrary nonempty and compact set $A \subset \mathbb{R}^d$ with $\lambda^d(A) > 0$

$$\begin{aligned}Q_r([0,1])^{1/r}\lambda^d(A)^{1/d} &= Q_r(A)^{1/r}\\ &= \lim_{n\to\infty} n^{1/d} e_{n,r}(U(A))\\ &\leq \liminf_{n\to\infty} n^{1/d} e_{n,\infty}(A)\\ &\leq \limsup_{n\to\infty} n^{1/d} e_{n,\infty}(A)\\ &\leq Q_\infty([0,1]^d)\lambda^d(A)^{1/d}, \ 1 \leq r < \infty,\end{aligned}$$

Here the first inequality follows from the Lemmas 10.1 and 10.6 (a) while the last inequality follows from (10.8). Since by assumption $\lim_{r\to\infty} Q_r([0,1]^d)^{1/r} = Q_\infty([0,1]^d)$, this implies

$$\lim_{n\to\infty} n^{1/d} e_{n,\infty}(A) = Q_\infty([0,1]^d)\lambda^d(A)^{1/d}.$$

10.9 Remark
Let $A \subset \mathbb{R}^d$ be an infinite compact set with $\lambda^d(\partial A) = 0$. For $\varepsilon > 0$, let $N(\varepsilon, A)$ be the minimal number of balls of radius $\varepsilon > 0$ which are necessary to cover A, i. e.,

$$\begin{aligned} N(\varepsilon) = N(\varepsilon, A) &= \min\{n \geq 1 \colon e_{n,\infty}(A) \leq \varepsilon\} \\ &= \min\Big\{n \geq 1 \colon \exists \alpha \subset \mathbb{R}^d,\ |\alpha| \leq n,\ \frac{1}{\varepsilon} A \subset \bigcup_{a \in \alpha} B(a, 1)\Big\}. \end{aligned}$$

Then

$$\varepsilon \geq e_{N(\varepsilon),\infty}(A),$$

hence by Theorem 10.7

$$\begin{aligned} \liminf_{\varepsilon \to 0} N(\varepsilon)\varepsilon^d &\geq \lim_{\varepsilon \to 0} N(\varepsilon) e_{N(\varepsilon),\infty}(A)^d \\ &= Q_\infty([0,1]^d)^d \lambda^d(A). \end{aligned}$$

The definition of $N(\varepsilon)$ implies

$$\varepsilon < e_{N(\varepsilon)-1,\infty}(A),$$

hence again by Theorem 10.7

$$\begin{aligned} \limsup_{\varepsilon \to 0} N(\varepsilon)\varepsilon^d &\leq \lim_{\varepsilon \to 0} N(\varepsilon) e_{N(\varepsilon)-1,\infty}(A)^d \\ &= Q_\infty([0,1]^d)^d \lambda^d(A). \end{aligned}$$

We obtain

$$\lim_{\varepsilon \to 0} N(\varepsilon, A)\varepsilon^d = Q_\infty([0,1]^d)^d \lambda^d(A). \tag{10.12}$$

(Actually, this limit result is equivalent to the assertion of Theorem 10.7). From (10.12) we deduce that $\lambda^d(B(0, Q_\infty([0,1]^d)))$ coincides with the density of the thinnest covering of the whole space by translates of $B(0,1)$ (cf. Gruber and Lekkerkerker, 1987, p. 237, Definition 6). The existence of the limit $\lim_{\varepsilon \to 0} N(\varepsilon, A)\varepsilon^d$ appears in Gruber and Lekkerkerker (1987, p. 237, Theorem 7) for convex compact bodies A.

10.3 Covering radius of lattices and bounds

As for the quantization coefficients, the covering coeffcient $Q_\infty([0,1]^d)$ is only known for $d = 1$, $d = 2$ (l_1-norm, l_2-norm), and in "trivial" cases for $d \geq 3$. Lower and upper bounds for $Q_\infty([0,1]^d)$ which correspond to those given in Proposition 8.3 and Theorem 8.9 can easily be derived.

For $A \subset \mathbb{R}^d$ compact with $\lambda^d(A) > 0$, set

$$M_\infty(A) = M_{1,\infty}(A).$$

Note that if $\Lambda \subset \mathbb{R}^d$ is an admissible lattice, then

$$(10.13) \qquad M_\infty(W(0|\Lambda)) = \frac{\max\{\|x\| : x \in W(0,\Lambda)\}}{\det(\Lambda)^{1/d}} = \frac{\sup_{x\in\mathbb{R}^d} \min_{a\in\Lambda} \|x-a\|}{\det(\Lambda)^{1/d}},$$

that is, $M_\infty(W(0|\Lambda))$ is the normalized covering radius of Λ (with respect to the whole space).

10.10 Proposition

(a) $\lim_{r\to\infty} Q_r([0,1]^d)^{1/r} \geq M_\infty(B(0,1)) = \dfrac{1}{\lambda^d(B(0,1))^{1/d}}.$

(b) If $\Lambda \subset \mathbb{R}^d$ is an admissible lattice then

$$Q_\infty([0,1]^d) \leq M_\infty(W(0|\Lambda)).$$

Proof

(a) By Proposition 8.3

$$\begin{aligned} \lim_{r\to\infty} Q_r([0,1]^d)^{1/r} &\geq \lim_{r\to\infty} M_r(B(0,1))^{1/r} \\ &= M_\infty(B(0,1)). \end{aligned}$$

(b) For $n \in \mathbb{N}$, let

$$\begin{aligned} c = c(n) &= (n \det(\Lambda))^{-1/d}, \\ \beta_{n,\Lambda} &= \{ca : a \in \Lambda,\ cW(a|\Lambda) \cap [0,1]^d \neq \emptyset\}, \\ k = k(n) &= |\beta_{n,\Lambda}| \end{aligned}$$

and

$$A_n = \bigcup_{b\in\beta_{n,\Lambda}} W(b|c\Lambda).$$

Then $1 \leq \lambda^d(A_n) = k/n$. If $U \subset \mathbb{R}^d$ is an open subset with $[0,1]^d \subset U$, then $A_n \subset U$ for sufficiently large n. Therefore

$$\begin{aligned} 1 &\leq \inf_{n\geq 1} \frac{k(n)}{n} = \inf_{n\geq 1} \lambda^d(A_n) \\ &\leq \inf\{\lambda^d(U) : U \subset \mathbb{R}^d \text{ open},\ [0,1]^d \subset U\} \\ &= \lambda^d([0,1]^d) = 1. \end{aligned}$$

Furthermore,

$$\begin{aligned} e_{k,\infty}([0,1]^d) &\le \max_{x\in[0,1]^d}\min_{b\in\beta_{n,\Lambda}} \|x-b\| \\ &\le \max_{x\in A_n}\min_{b\in\beta_{n,\Lambda}} \|x-b\| \\ &= \max\{\|x\| : x \in W(0|c\Lambda)\} \\ &= c\max\{\|x\| : x \in W(0|\Lambda)\} \\ &= n^{-1/d} M_\infty(W(0|\Lambda)). \end{aligned}$$

This implies

$$\begin{aligned} Q_\infty([0,1]^d) &= \inf_{n\ge 1} k^{1/d} e_{k,\infty}([0,1]^d) \\ &\le \inf_{n\ge 1} \left(\frac{k(n)}{n}\right)^{1/d} M_\infty(W(0|\Lambda)) \\ &= M_\infty(W(0|\Lambda)). \end{aligned}$$

□

Let the **lattice covering coefficient** of $[0,1]^d$ be defined by

$$Q_\infty^{(L)}([0,1]^d) = \inf\{M_\infty(W(0|\Lambda)) : \Lambda \subset \mathbb{R}^d \text{ admissible lattice}\}. \tag{10.14}$$

Then

$$Q_\infty([0,1]^d) \le Q_\infty^{(L)}([0,1]^d). \tag{10.15}$$

Note that $Q_r^{(L)}([0,1]^d)^{1/r}$ is increasing in r and

$$\lim_{r\to\infty} Q_r^{(L)}([0,1]^d)^{1/r} \le Q_\infty^{(L)}([0,1]^d). \tag{10.16}$$

If the underlying norm is the l_2-norm, $Q_\infty([0,1]^d)$ is known for $d=1$ and $d=2$. For $d=2$ and the hexagonal lattice $\Lambda \subset \mathbb{R}^2$, it follows from Theorem 8.15 that

$$\begin{aligned} \lim_{r\to\infty} Q_r([0,1]^2)^{1/r} &= \lim_{r\to\infty} M_r(W(0|\Lambda))^{1/r} \\ &= M_\infty(W(0|\Lambda)) \end{aligned}$$

and hence, by (10.11) and Proposition 10.10 (b)

$$\begin{aligned} Q_\infty([0,1]^2) &= \lim_{r\to\infty} Q_r([0,1]^2)^{1/r} \\ &= M_\infty(W(0|\Lambda)) \\ &= \left(\frac{2}{3\sqrt{3}}\right)^{1/2} = 0.6204\ldots, \quad l_2\text{-norm} \end{aligned} \tag{10.17}$$

This result is due to Kersher (1939). Solutions of the lattice covering problem are known for dimensions $1 \le d \le 5$ among them the hexagonal lattice for $d = 2$ and D_3^*. We have

(10.18)
$$\begin{aligned} Q_\infty^{(L)}([0,1]^d) &= (d+1)^{1/2d}\Big(\frac{d(d+2)}{12(d+1)}\Big)^{1/2}, \ 1 \le d \le 5, \ l_2\text{-norm} \\ Q_\infty^{(L)}([0,1]^3) &= \frac{2^{1/3}5^{1/2}}{4} = 0.7043\ldots, \\ Q_\infty^{(L)}([0,1]^4) &= \frac{2^{1/2}}{5^{3/8}} = 0.7733\ldots, \\ Q_\infty^{(L)}([0,1]^5) &= \frac{35^{1/2}}{2^{1/2}6^{9/10}} = 0.8340\ldots \end{aligned}$$

(cf. Conway and Sloane, 1993, p. 12 and Chapter 2, Section 1.3 and the subsequent Remark 10.10 (a)).

A trivial case occurs for the l_1-norm and $d = 2$, where $B(0,1) = W(0|D_2)$ (cf. Example 8.13). Therefore

(10.19)
$$\begin{aligned} Q_\infty([0,1]^2) &= \lim_{r\to\infty} Q_r([0,1]^2)^{1/r} \\ &= M_\infty(B(0,1)) = \frac{1}{\sqrt{2}} = 0.7071\ldots, \ l_1\text{-norm}. \end{aligned}$$

A further trivial case concerns the l_∞-norm. Then $B(0,1) = W(0|\mathbb{Z}^d)$ and so

(10.20)
$$\begin{aligned} Q_\infty([0,1]^d) &= \lim_{r\to\infty} Q_r([0,1]^d)^{1/r} \\ &= M_\infty(B(0,1)) = \frac{1}{2}, \ l_\infty\text{-norm} \end{aligned}$$

(cf. (8.8) and (8.9)).

10.11 Remark

(a) The lattice covering coefficient $Q_\infty^{(L)}([0,1]^d)$ coincides with the so called lower absolute inhomogenious minimum of the ball $B(0,1)$ and $\lambda^d\big(B(0, Q_\infty^{(L)}([0,1]^d))\big)$ coincides with the density of the thinnest lattice covering of the whole space with $B(0,1)$ provided every lattice is admissible (cf. Gruber and Lekkerkerker, 1987, p. 230 and p. 236). Note that

$$Q_\infty^{(L)}([0,1]^d) = \frac{1}{\sup\ \det(\Lambda)^{1/d}},$$

where the supremum is taken over all admissible lattices Λ such that $\{B(a,1) : a \in \Lambda\}$ is a covering of $\mathbb{R}^d$. For arbitrary (not necessarily admissible) lattices Λ one can show that

$$Q_\infty([0,1]^d) \le \frac{\max\{\|x\| : x \in W(0|\Lambda)\}}{\det(\Lambda)^{1/d}}$$

(cf. Gruber and Lekkerkerker, 1987, Theorem 6, p. 235).

(b) **(d-asymptotics)** If the underlying norm is the l_p-norm, with $1 \le p < \infty$, then

$$\liminf_{d\to\infty} d^{-1/p} Q_\infty([0,1]^d) \ge \frac{p}{2(ep)^{1/p}\Gamma(\frac{1}{p})}.$$

This follows from Proposition 9.4. The upper bound

$$Q_\infty([0,1]^d) \le \frac{(d \log d + d \log \log d + 5d)^{1/d}}{\lambda^d(B(0,1))^{1/d}}, \; d \ge 3$$

which is due to Rogers (1957) gives

$$\limsup_{d\to\infty} d^{-1/p} Q_\infty([0,1]^d) \le \frac{p}{2(ep)^{1/p}\Gamma(1/p)}.$$

Therefore

$$\lim_{d\to\infty} d^{-1/p} Q_\infty([0,1]^d) = \frac{p}{2(ep)^{1/p}\Gamma(1/p)}, \; l_p\text{-norm.} \tag{10.21}$$

Thus we find for the covering coefficient $Q_\infty([0,1]^d)$ exactly the same d-asymptotics as for the r-th roots $Q_r([0,1]^d)^{1/r}$ of the r-th quantization coefficients, $1 \le r < \infty$ (cf. Proposition 9.4).

(c) Let the underlying norm be the l_2-norm. For $d = 2$, the solution of the lattice quantizer problem — given by the hexagonal lattice — does not depend on r (cf. Theorem 8.15). For $d = 3$, the lattice D_3^* solves the lattice quantizer problem for $r = 2$ (cf. (8.8)) and the lattice covering problem ($r = \infty$). So possibly D_3^* solves the lattice quantizer problem for every r. For $d = 4$, the solution of the lattice covering problem is unique up to linear similarity transformations (cf. Baranovskii, 1965) and differs from the best known lattice quantizer D_4 for $r = 2$ (cf. Conway and Sloane, 1993, p. 12 and p. 61). Therefore, solutions of the lattice quantizer problem must depend on r.

One may use n-optimal (or asymptotically n-optimal) sets of centers of order $r = \infty$ as quantizers. Their asymptotic performance depends on the covering density of the unit ball (cf. Remark 10.8).

10.12 Proposition

Let $A \subset \mathbb{R}^d$ be a nonempty compact set with $\lambda^d(A) > 0$, let $(\alpha_n)_{n\ge1}$ be an asymptotically n-optimal set of centers for A of order ∞, i.e., $|\alpha_n| \le n$ and

$$\lim_{n\to\infty} n^{1/d} \max_{x\in A} \min_{a\in\alpha_n} \|x-a\| = Q_\infty([0,1]^d)\lambda^d(A)^{1/d},$$

and let $1 \le r < \infty$. Then

$$\limsup_{n\to\infty} n^{r/d} \int \min_{a\in\alpha_n} \|x-a\|^r \, dU(A)(x) \le \vartheta_d^{(d+r)/d} M_r(B(0,1))\lambda^d(A)^{r/d},$$

where $\vartheta_d = \lambda^d(B(0, Q_\infty([0,1]^d)))$. In particular

$$Q_r([0,1]^d) \le \vartheta_d^{(d+r)/d} M_r(B(0,1)).$$

Proof

Let $s_n = \max_{x \in A} d(x, \alpha_n)$. Then

$$s_n = \max_{a \in \alpha_n} \max_{x \in W(a|\alpha_n) \cap A} \|x - a\|$$

which gives

$$W(a|\alpha_n) \cap A \subset B(a, s_n), \; a \in \alpha_n.$$

This implies

$$\begin{aligned}
&\int \min_{a \in \alpha_n} \|x - a\|^r \, dU(A)(x) \\
&\leq \sum_{a \in \alpha_n} \int_{W(a|\alpha_n) \cap A} \|x - a\|^r \, dx / \lambda^d(A) \\
&\leq \sum_{a \in \alpha_n} \int_{B(a, s_n)} \|x - a\|^r \, dx / \lambda^d(A) \\
&\leq n M_r(B(0,1)) \lambda^d(B(0, s_n))^{(d+r)/d} / \lambda^d(A) \\
&= n s_n^{d+r} \lambda^d(B(0,1))^{(d+r)/d} M_r(B(0,1)) / \lambda^d(A).
\end{aligned}$$

Therefore

$$\begin{aligned}
&n^{r/d} \int \min_{a \in \alpha_n} \|x - a\|^r \, dU(A)(x) \\
&\leq (n s_n^d \lambda^d(B(0,1)))^{(d+r)/d} M_r(B(0,1)) / \lambda^d(A)
\end{aligned}$$

for every $n \in \mathbb{N}$. This yields the assertion. □

The above covering density upper bound for $Q_r([0,1]^d)$ is better than the "trivial" bound $Q_\infty([0,1]^d)^r$ as long as $\vartheta_d < (d+r)/d$ while for the r-th root

$$\lim_{r \to \infty} (\vartheta_d^{(d+r)/d} M_r(B(0,1)))^{1/r} = Q_\infty([0,1]^d).$$

10.4 Stability properties and empirical versions

A stability property for the n-th covering radius in terms of the metric d_H follows immediately from Lemma 10.4. If $A, B \subset \mathbb{R}^d$ are nonempty compact sets, then

$$|e_{n,\infty}(A) - e_{n,\infty}(B)| \leq d_H(A, B) \tag{10.22}$$

for every $n \in \mathbb{N}$. A stability result for n-optimal sets of centers of order $r = \infty$ can be derived from Lemma 4.22.

10.13 Theorem

Let $d_H(A_k, A) \to 0$ for nonempty compact sets $A_k, A \subset \mathbb{R}^d$ and let $\alpha_k \in C_{n,\infty}(A_k)$, $k \in \mathbb{N}$. Suppose

$$e_{n,\infty}(A) < e_{n-1,\infty}(A). \tag{10.23}$$

Then the set of d_H-cluster points of the sequence $(\alpha_k)_{k\geq 1}$ is a nonempty subset of $C_{n,\infty}(A)$ and

$$d_H(\alpha_k, C_{n,\infty}(A)) \to 0 \text{ as } k \to \infty.$$

Proof

To show that the asserton follows from Lemma 4.22 applied to the space of all nonempty compact subsets of $\mathbb{R}^d$ equipped with the Hausdorff metric d_H, the subset $N = \{\alpha \subset \mathbb{R}^d : |\alpha| \leq n, \alpha \neq \emptyset\}$, and $f = d_H(A, \cdot)$, it suffices to verify that

$$L(c) = \{\alpha \in N : d_H(\alpha, A) \leq c\}$$

is d_H-compact for some $c > e_{n,\infty}(A)$. By Lemma 10.4, this setting meets the covering problem because the assumptions imply $e_{n,\infty}(A_k) < e_{n-1,\infty}(A_k)$ for all large k. Choose $s > 0$ such that $A \subset B(0, s)$. Then

$$L(c) \subset \{\alpha \in N : \alpha \in B(0, c + s)\}.$$

Using Lemma 4.23 we deduce the d_H-compactness of $L(c)$. □

If in the preceding theorem the sets $\alpha_k \in C_{n,\infty}(A_k)$ satisfy $\max_{a\in\alpha_k} d(a, A_k) \leq e_{n,\infty}(A_k)$ for all large k (such a choice is always possible), then the assumption (10.23) can be dropped.

Under suitable conditions, weak convergence of probability distributions implies the d_H-convergence of their (compact) supports. The following special case will be needed.

10.14 Lemma
Let $P_k \xrightarrow{D} P$ for Borel probability measures on $\mathbb{R}^d$ with compact supports A_k and A, respectively. Then

$$\max_{x\in A} \min_{y\in A_k} \|x - y\| \to 0, \; k \to \infty.$$

Hence, if $A_k \subset A$ for every $k \in \mathbb{N}$, then

$$\lim_{k\to\infty} d_H(A_k, A) = 0.$$

Proof

For $\varepsilon > 0$, choose a finite subset α of A such that $A \subset \bigcup_{a\in\alpha} B(a, \varepsilon)$. For $a \in \alpha$, define a bounded continuous function $f_a : \mathbb{R}^d \to \mathbb{R}_+$ by

$$f_a(x) = \max\{0, 1 - \|x - a\|/\varepsilon\}.$$

Then

$$\max_{a\in\alpha} \left| \int f_a \, dP_k - \int f_a \, dP \right| \to 0, \; k \to \infty.$$

Since $\min_{a\in\alpha} \int f_a \, dP > 0$, one gets

$$\min_{a\in\alpha} P_k(B(a, \varepsilon)) \geq \min_{a\in\alpha} \int f_a \, dP_k > 0$$

and therefore

$$\max_{a\in\alpha}\min_{y\in A_k}\|a-y\|\leq\varepsilon$$

for sufficiently large k. This implies the assertion. □

From the stability properties one immediately obtains consistency results for empirical versions of the covering problem. Let $X_1, X_2, \ldots$ be i.i.d. $\mathbb{R}^d$–valued random variables with distribution P. The empirical version of $e_{n,\infty}(P)$ is given by

$$e_{n,\infty}(P_k)=e_{n,\infty}(\{X_1,\ldots,X_k\})=\inf_{|\alpha|\leq n}\max_{1\leq i\leq k}\min_{a\in\alpha}\|X_i-a\|,$$

where P_k denotes the empirical measure of $X_1, \ldots, X_k$.

10.15 Corollary (Consistency)
Let A denote the support of P and suppose A is compact.

(a) $d_H(\{X_1,\ldots,X_k\},A)=\max_{y\in A}\min_{1\leq i\leq k}\|y-X_i\|\to 0$ a. s., $k\to\infty$.

(b) $e_{n,\infty}(\{X_1,\ldots,X_k\})\to e_{n,\infty}(A)$ *a. s. as $k\to\infty$ uniformly in n.*

(c) Let $\alpha_k=\alpha_k(X_1,\ldots,X_k)\in C_{n,\infty}(\{X_1,\ldots,X_k\})$, $k\in\mathbb{N}$. Suppose (10.23) for $A=\operatorname{supp}(P)$. Then

$$d_H(\alpha_k,C_{n,\infty}(A))\to 0 \text{ a. s., } k\to\infty.$$

Proof

Since $P_k\xrightarrow{D}P$ a. s., the assertions follow from Theorem 10.13, Lemma 10.14, and (10.22). □

Notice that uniqueness $|C_{n,\infty}(A)|=1$ implies (10.23). Under this uniqueness condition, Corollary 10.15 (c) is contained in Cuesta-Albertos et al. (1988, Theorem 12). Part(a) has been observed by Wagner (1971).

Notes

Some material about the issue of this section for the l_2-norm and l_∞-norm may be found in Niederreiter (1992), Chapter 6. In particular, the exact order $n^{-1/d}$ of $e_{n,\infty}(A)$ is well known if $\lambda^d(A)>0$. However, we are not aware of a reference concerning Theorem 10.7. Examples of n–optimal sets of centers for $[0,1]^2$ of order ∞ can be found in Johnson et al. (1990) for the l_1–norm and the l_2–norm. The covering density upper bound for the r-th quantization coefficients given in Proposition 10.12 seems to be new. A discussion of the relation between the quantization problem for $r=2$, the covering problem and the packing problem can be found in Forney (1993) for the l_2-norm. For general treatments of the covering problem we refer to Gruber and Lekkerkerker (1987) and Conway and Sloane (1993).

Consistency and central limit results for a trimmed version of the covering problem have been proved by Cuesta-Albertos et al. (1998) and Cuesta–Albertos et al. (1999). Empirical versions of related covering problems and their asymptotics when both the level n and the sample size k tend to infinity were studied e. g. by Zemel (1985) and Rhee and Talagrand (1989b) for the l_2-norm.

Let us mention that n-optimal sets of centers of order ∞ are often called best n-nets and Chebyshev-centers in case $n = 1$. (cf. Garkavi, 1964, and Singer, 1970, Section II.6.4).

10.16 Conjecture
$\lim_{r\to\infty} Q_r([0,1]^d)^{1/r} = Q_\infty([0,1]^d)$ (cf. (10.11) and Remark 10.8).

If Conjecture 8.17 can be resolved, then Conjecture 10.16 is true for $d = 3$ and l_2–norm. Furthermore, if Conjecture 8.17 can be resolved, then the lattice D_3^* provides a solution of the covering problem in $\mathbb{R}^3$ for the l_2–norm, i.e., $Q_\infty([0,1]^3) = M_\infty(W(0|D_3^*))$. This is a long standing conjecture in geometry.

Chapter III

Asymptotic quantization for singular probability distributions

In this chapter we consider some classes of continuous singular distributions on $\mathbb{R}^d$ and determine the asymptotic first order behaviour of their quantization errors.

11 The quantization dimension

Here we determine the order of convergence for the sequence of quantization errors of a given distribution. X is an $\mathbb{R}^d$-valued random variable and P is its distribution. In some cases we abbreviate $e_{n,r}(P)$ by $e_{n,r}$, $V_{n,r}(P)$ by $V_{n,r}$, and $C_{n,r}(P)$ by $C_{n,r}$. In this section we always assume either that $1 \leq r < \infty$ and $E(\|X\|^r) < +\infty$ or that $r = \infty$ and $\mathrm{supp}(P)$ is compact.

11.1 Definition and elementary properties

11.1 Definition
$\underline{D}_r := \underline{D}_r(P) = \liminf_{n\to\infty} \frac{\log n}{-\log e_{n,r}}$ is called the **lower quantization dimension of P of order r**.

$\overline{D}_r := \overline{D}_r(P) = \limsup_{n\to\infty} \frac{\log n}{-\log e_{n,r}}$ is called the **upper quantization dimension of P of order r**.

If the two numbers $\underline{D}_r$ and $\overline{D}_r$ agree then their common value is denoted by D_r $(= D_r(P))$ and called the **quantization dimension of P of order r**.

11.2 Remark
$\underline{D}_r$ and $\overline{D}_r$ do not depend on the underlying norm. $\underline{D}_\infty$ and $\overline{D}_\infty$ depend only on the support of P. Using the definition of $e_{n,\infty}(K)$ in (10.3) we also define $\underline{D}_\infty(K)$, $\overline{D}_\infty(K)$, and $D_\infty(K)$ for an arbitrary (nonempty) compact set $K \subset \mathbb{R}^d$.

11.3 Proposition

(a) If $0 \leq t < \underline{D}_r < s$ *then*

$$\lim_{n\to\infty} ne_{n,r}^t = +\infty \text{ and } \liminf_{n\to\infty} ne_{n,r}^s = 0$$

(b) If $0 \leq t < \overline{D}_r < s$ *then*

$$\limsup_{n\to\infty} ne_{n,r}^t = +\infty \text{ and } \lim_{n\to\infty} ne_{n,r}^s = 0.$$

Proof

Let us first prove (a). If $e_{n,r} = 0$ for some $n \in \mathbb{N}$ then $\underline{D}_r = 0$ and (a) is obvious. Suppose $e_{n,r} > 0$ for all $n \in \mathbb{N}$. For $0 \leq t < \underline{D}_r$ choose $t' \in (t, \underline{D}_r)$. Then there exists an $n_0 \in \mathbb{N}$ with

$$e_{n,r} < 1 \text{ and } \frac{\log n}{-\log e_{n,r}} > t'$$

for all $n \geq n_0$. This implies

$$ne_{n,r}^{t'} > 1$$

and, hence

$$ne_{n,r}^t > e_{n,r}^{t-t'}$$

for all $n \geq n_0$. Since $\lim_{n\to\infty} e_{n,r} = 0$ we deduce

$$\lim_{n\to\infty} ne_{n,r}^t = +\infty.$$

For $\underline{D}_r < s$ there is an $s' \in (\underline{D}_r, s)$ and a subsequence $(e_{n_k,r})$ of $(e_{n,r})$ with

$$e_{n_k,r} \leq 1 \text{ and } \frac{\log n_k}{-\log e_{n_k,r}} \leq s'.$$

for all $k \in \mathbb{N}$. This implies

$$n_k e_{n_k,r}^{s'} \leq 1$$

and, hence

$$n_k e_{n_k,r}^s \leq e_{n_k,r}^{s-s'}$$

Since $\lim_{n\to\infty} e_{n_k,r} = 0$ this leads to

$$\liminf_{n\to\infty} ne_{n,r}^s \leq \lim_{k\to\infty} n_k e_{n_k,r}^s = 0.$$

Part (b) can be proved in a similar way. □

11.4 Corollary

(a) If $1 \leq r \leq s \leq \infty$ *then* $\underline{D}_r \leq \underline{D}_s$ *and* $\overline{D}_r \leq \overline{D}_s$.

(b) *If $D \in (0, +\infty)$ is such that*

$$0 < \liminf_{n\to\infty} n e_{n,r}^D \le \limsup_{n\to\infty} n e_{n,r}^D < +\infty$$

then $D_r = D$.

(c) *Let $1 \le r < \infty$ and suppose $E(\|X\|^{r+\delta}) < +\infty$ for some $\delta > 0$ then $\overline{D}_r \le d$. If the absolutely continuous part P_a of P does not vanish then $D_r = d$.*

(d) *Let $r = \infty$. Then $\overline{D}_\infty \le d$. If $\lambda^d(\mathrm{supp}(P)) > 0$ then $D_\infty = d$.*

Proof

(a) follows from Lemma 10.1 (a) and Proposition 11.3
(b) follows immediately from Proposition 11.3
(c) and (d) follow from Proposition 11.3, Theorem 6.2, (10.8), and the fact that

$$n e_{n,\infty}(P)^d \ge \lambda^d(\mathrm{supp}(P))/\lambda^d(B(0,1)).$$

□

11.2 Comparison to the Hausdorff dimension

Next we will investigate the connection of the quantization dimension to other types of dimension. Let us first consider the relationship of $\underline{D}_\infty(P)$ to the Hausdorff dimension of the support of P.
For a set $A \subset \mathbb{R}^d$ and $\varepsilon > 0$ an ε-cover of A is a cover of A by sets U_i each of diameter at most ε, i. e., $\mathrm{diam}(U_i) = \sup\{\|x - y\| : x, y \in A\} \le \varepsilon$. For $s \ge 0$ let

$$\mathcal{H}_\varepsilon^s(A) = \inf\{\sum_{i\in I} \mathrm{diam}(U_i)^s : (U_i)_{i\in I} \text{ is an } \varepsilon\text{-cover of } A\}$$

Then $\mathcal{H}^s(A) = \lim_{\varepsilon\to 0} \mathcal{H}_\varepsilon^s(A)$ is the **s-dimensional Hausdorff measure** of A. It is easy to check that $\mathcal{H}^s(A)$ is non-increasing with s and that $\mathcal{H}^t(A) > 0$ implies $\mathcal{H}^s(A) = \infty$ for all $s < t$. The **Hausdorff dimension** of A is defined as

$$\begin{aligned}\dim_H(A) &= \sup\{s \ge 0 : \mathcal{H}^s(A) = \infty\} \\ &= \inf\{s \ge 0 : \mathcal{H}^s(A) = 0\}.\end{aligned}$$

While the definition of the Hausdorff measure depends on the underlying norm the Hausdorff dimension has the same value for all norms on $\mathbb{R}^d$.

11.5 Proposition
Let $K \subset \mathbb{R}^d$ be compact and let P be any probability measure with $\mathrm{supp}(P) = K$. Then, for every $t \ge 0$,

$$\mathcal{H}^t(K) \le 2^t \liminf_{n\to\infty} n e_{n,\infty}(P)^t, \tag{11.1}$$

in particular

$$\dim_H(K) \le \underline{D}_\infty(P).$$

Proof

Let $\alpha_n \in C_{n,\infty}$. Then $(B(a, e_{n,\infty}))_{a\in\alpha_n}$ is a cover of K. Let $\varepsilon > 0$ be arbitrary. Since $(e_{n,\infty})_{n\in\mathbb{N}}$ converges to 0 there is an $n_\varepsilon \in \mathbb{N}$ with $e_{n,\infty} \leq \varepsilon$ for all $n \geq n_\varepsilon$. This implies

$$\begin{aligned}\mathcal{H}^t_\varepsilon(K) &\leq \inf_{n\geq n_\varepsilon} \sum_{a\in\alpha_n} \operatorname{diam}(B(a, e_{n,\infty}))^t \\ &\leq \inf_{n\geq n_\varepsilon} n(2e_{n,\infty})^t\end{aligned}$$

By letting ε tend to 0 we obtain

$$\mathcal{H}^t(K) \leq 2^t \liminf_{n\to\infty} ne^t_{n,\infty}.$$

The remaining claim in the proposition follows from Proposition 11.3. □

The **Hausdorff dimension of a** (probability) **measure** P is defined to be

(11.2) $$\dim_H(P) = \inf\{\dim_H(A)\colon A \in \mathcal{B}(\mathbb{R}^d), P(\mathbb{R}^d\backslash A) = 0\}.$$

11.6 Theorem
For all $r \geq 1$,

$$\dim_H(P) \leq \underline{D}_r(P).$$

Proof.
The proof will be given in Corollary 12.16. □

Since $\dim_H(P) \leq d$, the above inequality can be strict (see Exmaple 6.4). In the case that $r = 2$ and $\operatorname{supp}(P)$ is compact the above theorem was proved by Pötzelberger(1998a).

11.3 Comparison to the box dimension

Now we will consider the relationship between $\underline{D}_\infty(P)$, $\overline{D}_\infty(P)$ and the upper and lower box dimension of $\operatorname{supp}(P)$.

For our purposes the box dimension (entropy dimension, Minkowski dimension) of a compact subset K of $\mathbb{R}^d$ is most conveniently defined in the following way:

For $\varepsilon > 0$ let

$$N(\varepsilon) = N(\varepsilon, K) = \min\{n \in \mathbb{N}\colon e_{n,\infty}(K) \leq \varepsilon\}.$$

Then

(11.3) $$\underline{\dim}_B(K) = \liminf_{\varepsilon\to 0} \frac{\log N(\varepsilon)}{-\log\varepsilon}$$

is called the **lower box dimension** of K and

(11.4) $$\overline{\dim}_B(K) = \limsup_{\varepsilon\to 0} \frac{\log N(\varepsilon)}{-\log\varepsilon}$$

is called the **upper box dimension** of K.

If $\underline{\dim}_B(K) = \overline{\dim}_B(K)$ this value is denoted by $\dim_B(K)$ and called the **box dimension** of K.

This definition suggests that there is a close relationship between $\underline{D}_\infty(K)$ and $\underline{\dim}_B(K)$ and between $\overline{D}_\infty(K)$ and $\overline{\dim}_B(K)$. We have the following result.

11.7 Theorem
Let $K \subset \mathbb{R}^d$ be compact. Then

(i) $\underline{\dim}_B(K) = \underline{D}_\infty(K)$,

(ii) $\overline{\dim}_B(K) = \overline{D}_\infty(K)$.

Proof

(i) To prove $\underline{\dim}_B(K) \leq \underline{D}_\infty(K)$ let $n \geq 1$ be a natural number. Then

$$N_n := N(e_{n,\infty}) \leq n \text{ and } e_{N_n,\infty} = e_{n,\infty}.$$

We deduce

$$\begin{aligned} \underline{D}_\infty(K) &= \liminf_{n\to\infty} \frac{\log n}{-\log e_{n,\infty}} \geq \liminf_{n\to\infty} \frac{\log N_n}{-\log e_{n,\infty}} \\ &\geq \liminf_{\varepsilon\to 0} \frac{\log N(\varepsilon)}{-\log \varepsilon} = \underline{\dim}_B(K). \end{aligned}$$

Next we show $\underline{\dim}_B(K) \geq \underline{D}_\infty(K)$. For $\varepsilon > 0$ the definition of $N(\varepsilon)$ implies

$$e_{N(\varepsilon),\infty} \leq \varepsilon,$$

hence

$$\begin{aligned} \underline{D}_\infty(K) &= \liminf_{n\to\infty} \frac{\log n}{-\log e_{n,\infty}} \leq \liminf_{\varepsilon\to 0} \frac{\log N(\varepsilon)}{-\log e_{N(\varepsilon),\infty}} \\ &\leq \liminf_{\varepsilon\to 0} \frac{\log N(\varepsilon)}{-\log \varepsilon} = \underline{\dim}_B(K). \end{aligned}$$

(ii) First we will prove that there is a $k \geq 1$ such that

$$e_{kn,\infty} \leq \tfrac{1}{2} e_{n,\infty}$$

for all $n \geq 1$. For $n \geq 1$ choose $\alpha \in C_{n,\infty}$. Let $\beta \subset \mathbb{R}^d$ be of minimum cardinality with

$$d_\beta(x) \leq \tfrac{1}{2}$$

for all $x \in B(0,1)$. Let $k = |\beta|$. For $y \in \mathbb{R}^d$, $\varepsilon > 0$ and $\beta(y,\varepsilon) = \varepsilon\beta + y$ we have

$$d_{\beta(y,\varepsilon)}(x) \leq \tfrac{1}{2}\varepsilon$$

for all $x \in B(y, \varepsilon)$. Let $\alpha' = \bigcup_{y \in \alpha} \beta(y, e_{n,\infty})$.
Then $|\alpha'| \leq kn$ and for every $x \in K$ there is a $y \in \alpha$ with $\|x - y\| \leq e_{n,\infty}$ and hence a $z \in \beta(y, e_{n,\infty})$ with

$$\|x - z\| \leq \tfrac{1}{2} e_{n,\infty}.$$

Thus we obtain $d(x, \alpha') \leq \frac{1}{2} e_{n,\infty}$.
This implies

$$e_{kn,\infty} \leq \sup_{x \in K} d_{\alpha'}(x) \leq \tfrac{1}{2} e_{n,\infty}.$$

Next we will show that, for $n \in \mathbb{N}$ with $e_{n,\infty} > 0$, we have

$$N(e_{n,\infty}) \leq n < kN(e_{n,\infty}). \tag{11.5}$$

Let $N_n = N(e_{n,\infty})$. Then $N_n \leq n$ and since $e_{N_n,\infty} = e_{n,\infty} > 0$, we have

$$e_{kN_n,\infty} \leq \tfrac{1}{2} e_{N_n,\infty} < e_{n,\infty}.$$

This implies $n < kN_n$.

Using (11.5) and $\lim_{n \to \infty} e_{n,\infty} = 0$ we get

$$\begin{aligned}
\overline{\dim}_B(K) &= \limsup_{\varepsilon \to 0} \frac{\log N(\varepsilon)}{-\log \varepsilon} \geq \limsup_{n \to \infty} \frac{\log N(e_{n,\infty})}{-\log e_{n,\infty}} \\
&\geq \limsup_{n \to \infty} \frac{\log \frac{1}{k} n}{-\log e_{n,\infty}} \\
&= \limsup_{n \to \infty} \left[\frac{-\log k}{-\log e_{n,\infty}} + \frac{\log n}{-\log e_{n,\infty}} \right] \\
&= \overline{D}_\infty(K).
\end{aligned}$$

To prove the converse inequality observe that, for small $\varepsilon > 0$,

$$e_{N(\varepsilon)-1,\infty} > \varepsilon.$$

This leads to

$$\begin{aligned}
\overline{\dim}_B(K) &= \limsup_{\varepsilon \to 0} \frac{\log N(\varepsilon)}{-\log \varepsilon} \\
&\leq \limsup_{\varepsilon \to 0} \frac{\log N(\varepsilon)}{-\log e_{N(\varepsilon)-1,\infty}} \\
&= \limsup_{\varepsilon \to 0} \frac{\log(N(\varepsilon) - 1)}{-\log e_{N(\varepsilon)-1,\infty}} \frac{\log N(\varepsilon)}{\log(N(\varepsilon) - 1)} \\
&= \limsup_{\varepsilon \to 0} \frac{\log(N(\varepsilon) - 1)}{-\log e_{N(\varepsilon)-1,\infty}} \\
&\leq \overline{D}_\infty
\end{aligned}$$

and completes the proof of the theorem. □

11.8 Corollary
Let $K \subset \mathbb{R}^d$ be compact. If the box dimension of K exists then the quantization dimension of K of order ∞ also exists and equals the box dimension.

11.9 Proposition
Let P be a probability on $\mathbb{R}^d$ with compact support K. Then, for $1 \leq r \leq s \leq \infty$

$$\overline{D}_r(P) \leq \overline{D}_s(P) \leq \overline{D}_\infty(P) = \overline{D}_\infty(K) = \overline{\dim}_B(K)$$

and

$$\underline{D}_r(P) \leq \underline{D}_s(P) \leq \underline{D}_\infty(P) = \underline{D}_\infty(K) = \underline{\dim}_B(K).$$

Proof

The result follows immediately from Corollary 11.4 and Theorem 11.7. □

11.4 Comparison to the rate distortion dimension

T. Kawabata and A. Dembo (1994) introduced the concept of rate distortion dimension for probability distributions. They showed that for norms on $\mathbb{R}^d$ the rate distortion dimension is the same as Rényi's information dimension. Here we will compare the quantization dimension to the rate distortion dimension.

Let us recall its definition. For $x = 0$ set $x \log(x) = 0$. For a probability Q on the Borel σ-field of $\mathbb{R}^d \times \mathbb{R}^d$ denote by Q_1 and Q_2 the marginals on the first and second component, respectively. If P is a probability on $\mathbb{R}^d$ and Q is a probability on $\mathbb{R}^d \times \mathbb{R}^d$ with $P = Q_1$ then the **average mutual information** $I(P, Q)$ of Q is equal to

$$\int h(x,y) \log h(x,y) \, dQ_1 \otimes Q_2(x,y)$$

if Q is absolutely continuous with respect to $Q_1 \otimes Q_2$ and h is the corresponding Radon-Nikodym derivative and equal to ∞ otherwise.
Let $1 \leq r < \infty$. The **rate distortion function of order r**, $R_{P,r} \colon (0, +\infty) \to \mathbb{R}$, is defined by

$$R_{P,r}(t) = \inf\{I(P,Q) \colon Q \text{ probability on } \mathbb{R}^d \times \mathbb{R}^d \text{ with } Q_1 = P \text{ and } \int \|x - y\|^r \, dQ(x,y) \leq t\}.$$

The **upper rate distortion dimension** (of order r) of P is defined to be

$$\overline{\dim}_R(P) = \limsup_{\varepsilon \to 0} \frac{R_{P,r}(\varepsilon^r)}{-\log \varepsilon}.$$

The **lower rate distortion dimension** (of order r) of P is

$$\underline{\dim}_R(P) = \liminf_{\varepsilon \to 0} \frac{R_{P,r}(\varepsilon^r)}{-\log \varepsilon}$$

and, if the two values agree, it is called the **rate distortion dimension** (of order r) of P and denoted by $\dim_R(P)$.
It is shown in Kawabata and Dembo (1994, Proposition 3.3) that the (upper, respectively lower) rate distortion dimension does not depend on r and equals the corresponding (upper, respectively lower) information dimension introduced by Rényi (1959).

11.10 Theorem
If $1 \leq r < \infty$ *then*

$$\underline{\dim}_R(P) \leq \underline{D}_r(P).$$

Proof

Let $\varepsilon > 0$ be given. Let $n \in \mathbb{N}$ satisfy $e_{n,r} \leq \varepsilon$. Let $f\colon \mathbb{R}^d \to \mathbb{R}^d$ be an n-optimal quantizer of order r and Q the image of P on $\mathbb{R}^d \times \mathbb{R}^d$ under the map $x \to (x, f(x))$. Then $Q_1 = P$, $Q_2 = P^f = P \circ f^{-1}$ and

$$V_{n,r} = \int \|x - f(x)\|^r \, dP(x) = \int \|x - y\|^r \, dQ(x,y) \leq \varepsilon^r.$$

Set $\alpha = f(\mathbb{R}^d)$ and define $h\colon \mathbb{R}^d \times \mathbb{R}^d \to \mathbb{R}$ by

$$h(x,y) = \begin{cases} 0, & y \notin \alpha \\ \frac{1}{P(f=y)} 1_{\{f=y\}}(x), & y \in \alpha. \end{cases}$$

For a Borel set $A \subset \mathbb{R}^d \times \mathbb{R}^d$ we obtain

$$\begin{aligned}
\int_A h(x,y) \, dQ_1 \otimes Q_2(x,y) &= \iint 1_A(x,y) h(x,y) \, dQ_2(y) \, dP(x) \\
&= \int \sum_{a \in \alpha} P(f=a) 1_A(x,a) h(x,a) \, dP(x) \\
&= \sum_{a \in \alpha} \int P(f=a) 1_A(x,a) \frac{1}{P(f=a)} 1_{\{f=a\}}(x) \, dP(x) \\
&= \sum_{a \in \alpha} \int_{\{f=a\}} 1_A(x,a) \, dP(x) \\
&= \sum_{a \in \alpha} P(\{x\colon f(x) = a \text{ and } (x, f(x)) \in A\}) \\
&= P(\{x\colon (x, f(x)) \in A\}) \\
&= Q(A).
\end{aligned}$$

Thus Q is absolutely continuous with respect to $Q_1 \otimes Q_2$ and h is the corresponding

Radon-Nikodym derivative. By the definition of $R_{P,r}$ we get

$$\begin{aligned}
R_{P,r}(\varepsilon^r) &\le I(P,Q) \\
&= \int h(x,y)\log h(x,y)\, dQ_1 \otimes Q_2(x,y) \\
&= \int \log h(x,y)\, dQ(x,y) \\
&= \int \log h(x,f(x))\, dP(x) \\
&= \sum_{a\in\alpha} \int_{\{f=a\}} \log h(x,a)\, dP(x) \\
&= \sum_{a\in\alpha} \int_{\{f=a\}} \log \frac{1}{P(f=a)}\, dP(x) \\
&= -\sum_{a\in\alpha} P(f=a)\log P(f=a) \\
&\le \log|\alpha|.
\end{aligned}$$

Since f is n-optimal and $\alpha = f(\mathbb{R}^d)$ we know that $|\alpha| = n$ (cf. Theorem 4.1). Thus we have shown that

$$e_{n,r} \le \varepsilon \text{ implies } R_{P,r}(\varepsilon^r) \le \log n.$$

Now let $n_\varepsilon \in \mathbb{N}$ be the smallest natural number with $e_{n_\varepsilon,r} \le \varepsilon$. Then we get

$$R_{P,r}(V_{n_\varepsilon,r}) \le \log n_\varepsilon$$

hence

$$\frac{R_{P,r}(V_{n_\varepsilon,r})}{-\log e_{n_\varepsilon,r}} \le \frac{\log n_\varepsilon}{-\log e_{n_\varepsilon,r}}.$$

This implies

$$\liminf_{\varepsilon\to 0} \frac{R_{P,r}(\varepsilon^r)}{-\log \varepsilon} \le \liminf_{\varepsilon\to 0} \frac{\log n_\varepsilon}{-\log e_{n_\varepsilon,\infty}}.$$

With $\varepsilon_n = e_{n,r}$ we know that $n_{\varepsilon_n} \le n$ and $e_{n_{\varepsilon_n},r} = e_{n,r}$ and get

$$\liminf_{\varepsilon\to\infty} \frac{\log n_\varepsilon}{-\log e_{n_\varepsilon,r}} \le \liminf_{n\to\infty} \frac{\log n}{-\log e_{n,r}} = \underline{D}_r(P).$$

Thus the theorem is proved. □

11.11 Remark

It follows from the preceding theorem that, for $1 \le r \le \infty$,

$$\underline{\dim}_R(P) \le \underline{D}_1(P) \le \underline{D}_r(P) \le \underline{D}_\infty(P) = \underline{\dim}_B(\mathrm{supp}(P)).$$

It remains an open question whether $\overline{\dim}_R(P) \le \overline{D}_1(P)$.

Notes

The concept of quantization dimension was introduced by Zador (1982). Hausdorff- and box dimension are classical mathematical notions which play a central role in fractal geometry. A good survey can be found in the books of Falconer (1985, 1990, 1997). The rate distortion dimension is introduced and thoroughly discussed by Kawabata and Dembo (1994) where the identity with Rényi's information dimension is also pointed out (cf. Rényi, 1959).

12 Regular sets and measures of dimension D

The notion of regular sets and measure of dimension D is an obvious modification of a concept of regularity used by David and Semmes (1993) and attributed to Ahlfors by these authors. The class of regular sets contains, for instance, convex sets in $\mathbb{R}^d$, surfaces of these sets, compact C^1-manifolds, and self–similar sets (satisfying the open set condition). Examples of regular measures of dimension D are certain measures which are absolutely continuous with respect to the Hausdorff measure on a regular set of dimension D. Here we give a detailed discussion of the asymptotic behaviour of the quantization errors for regular measures of dimension D. In this section $\| \ \|$ is an arbitrary norm on $\mathbb{R}^d$ and D is a nonnegative real number. By $\overset{\circ}{B}(a,r)$ we denote the open ball of center a and radius r.

12.1 Definition and examples

12.1 Definition

Let μ be a finite Borel measure on $\mathbb{R}^d$.

(a) μ is called **regular of dimension D** if μ has compact support and satisfies

$$\exists c > 0 \, \exists r_0 > 0 \, \forall r \in (0, r_0) \, \forall a \in \operatorname{supp}(\mu) : \tfrac{1}{c} r^D \le \mu(\overset{\circ}{B}(a,r)) \le c r^D.$$

(b) $M \subset \mathbb{R}^d$ is called **regular of dimension D** if M is compact, $0 < \mathcal{H}^D(M) < \infty$, and the restriction $\mathcal{H}^D_{|M} = \mathcal{H}^D(\cdot \cap M)$ of $\mathcal{H}^D$ to M is a regular measure of dimension D with support M.

12.2 Remark

A set or a measure which is regular of dimension D in $\mathbb{R}^d$ with one given norm is also regular with dimension D in $\mathbb{R}^d$ with any other norm. This follows from the wellknown fact that any two norms on $\mathbb{R}^d$ are equivalent, i. e., if $\| \ \|$ and $||| \ |||$ are norms on $\mathbb{R}^d$ then there is a constant $c > 0$ with $\frac{1}{c} ||| \ ||| \le \| \ \| \le c ||| \ |||$. The notion of regularity of dimension D remains unchanged if one uses closed balls instead of open balls in the definition.

Next we will study the elementary properties of regular sets of dimension D.

12.3 Lemma

Let μ be a finite measure on $\mathbb{R}^d$ such that there is a $c > 0$ and an $r_0 > 0$ with $\mu(\overset{\circ}{B}(a,r)) \le c r^D$ for all $a \in \operatorname{supp}(\mu)$ and all $r \in (0, r_0)$. Then there is a $c' > 0$ with

$$\mu(\overset{\circ}{B}(a,r)) \le c' r^D \quad \text{for all } a \in \mathbb{R}^d \text{ and all } r > 0.$$

Proof

First we will show that there is a $\tilde{c} > 0$ with $\mu(\mathring{B}(a,r)) \leq \tilde{c}r^D$ for all $a \in \text{supp}(\mu)$ and all $r > 0$. To this end let $a \in \text{supp}(\mu)$ be arbitrary and define

$$\tilde{c} = \max(c, \mu(\mathbb{R}^d)r_0^{-D}).$$

If $r \in (0, r_0)$ then by assumption we have

$$\mu(\mathring{B}(a,r)) \leq cr^D \leq \tilde{c}r^D.$$

If $r \geq r_0$ then

$$\tilde{c}r^D \geq \mu(\mathbb{R}^d)r_0^{-D}r^D \geq \mu(\mathbb{R}^d) \geq \mu(\mathring{B}(a,r)).$$

We claim that, for arbitrary $a \in \mathbb{R}^d$ and $r > 0$,

$$\mu(\mathring{B}(a,r)) \leq 2^D\tilde{c}r^D = c'r^D.$$

If $0 < r \leq d(a, \text{supp}(\mu))$ then $\mu(\mathring{B}(a,r)) = 0$ and the claim is true.
If $d(a, \text{supp}(\mu)) < r$ choose $b \in \text{supp}(\mu)$ with $\|a - b\| = d(a, \text{supp}(\mu))$. Then we have

$$\begin{aligned}\mu(\mathring{B}(a,r)) &\leq \mu(\mathring{B}(b, r + \|b - a\|)) \\ &\leq \tilde{c}(r + \|b - a\|)^D \\ &\leq \tilde{c}r^D(1 + \frac{\|b - a\|}{r})^D \leq 2^D\tilde{c}r^D.\end{aligned}$$

□

12.4 Lemma
A finite union of regular sets of dimension D is regular of dimension D.

Proof

Let $M_1, \dots, M_n \subset \mathbb{R}^d$ be regular of dimension D and $M = M_1 \cup \dots \cup M_n$. Then M is compact and we have

$$0 < \mathcal{H}^D(M) \leq \sum_{i=1}^{n} \mathcal{H}^D(M_i) < \infty.$$

Let $c_i > 0$, $r_{i,0} > 0$ be such that

$$\forall r \in (0, r_{i,0}) \; \forall a \in M_i : \frac{1}{c_i}r^D \leq \mathcal{H}^D(M_i \cap \mathring{B}(a,r)) \leq c_i r^D.$$

By Lemma 12.3 there is a constant $c_i' > 0$ with

$$\mathcal{H}^D(M_i \cap \mathring{B}(a,r)) \leq c_i' r^D$$

for all $a \in \mathbb{R}^d$ and all $r > 0$. Without loss of generality we may assume $c_i' \geq c_i$. Set $c = c_1' + \ldots + c_n'$ and $r_0 = \min(r_{1,0}, \ldots, r_{n,0})$. It follows that for all $a \in M$ and all $r \in (0, r_0)$

$$\begin{aligned} \frac{1}{c} r \leq \min(\frac{1}{c_1}, \ldots, \frac{1}{c_n}) r^D &\leq \mathcal{H}^D(M \cap \mathring{B}(a,r)) \\ &\leq \sum_{i=1}^{n} \mathcal{H}^D(M_i \cap \mathring{B}(a,r)) \\ &\leq \sum_{i=1}^{n} c_i' r^D \\ &\leq c r^D. \end{aligned}$$

□

12.5 Lemma
Let $M \subset \mathbb{R}^d$ be compact. Then M is regular of dimension D if and only if every point x of M has a regular neighbourhood of dimension D in M.

Proof

Since M is compact M can be covered by finitely many regular sets of dimension D and the lemma follows from Lemma 12.4. □

12.6 Lemma
Let $M \subset \mathbb{R}^d$ be regular of dimension D, $U \subset \mathbb{R}^d$ open with $M \subset U$, and $g: U \to \mathbb{R}^d$ a bi-Lipschitz map, i. e., there is a constant $c' > 0$ with

$$\frac{1}{c'} \|x - y\| \leq \|g(x) - g(y)\| \leq c' \|x - y\|$$

for all $x, y \in U$.
Then $g(M)$ is regular of dimension D.

Proof

Obviously $g(U)$ is open and $g(M)$ is a compact subset of $g(U)$. It follows (from Falconer (1990, p. 28, 2.9)) that

$$0 < (\frac{1}{c'})^D \mathcal{H}^D(M) \leq \mathcal{H}^D(g(M)) \leq c'^D \mathcal{H}^D(M) < \infty.$$

Let $r_1 = \min_{z \in g(M)} d(z, g(U)^c)$. Then we have $r_1 > 0$. Let $r_0 > 0$ and $c > 0$ be such that

$$\frac{1}{c} r^D \leq \mathcal{H}^D(M \cap \mathring{B}(a,r)) \leq c r^D$$

for all $a \in M$ and $r \in (0, r_0)$. Define $r_0' = \min(r_1, \frac{1}{c'} r_0)$. For $y \in g(M)$ and $r \in (0, r_0)$ we obtain

$$\mathring{B}(g^{-1}(y), \frac{1}{c'} r) \subset g^{-1}(\mathring{B}(y,r)) \subset \mathring{B}(g^{-1}(y), c' r)$$

and, hence,

$$
\begin{aligned}
(12.1) \qquad \frac{1}{c}\left(\frac{1}{c'}r\right)^D &\le \mathcal{H}^D(M \cap \overset{\circ}{B}(g^{-1}(y), \frac{1}{c'}r)) \\
&\le \mathcal{H}^D(M \cap g^{-1}(\overset{\circ}{B}(y,r))) \\
&\le \mathcal{H}^D(M \cap \overset{\circ}{B}(g^{-1}(y), c'r)) \\
&\le c(c'r)^D
\end{aligned}
$$

Since $g(M \cap g^{-1}(\overset{\circ}{B}(y,r))) = g(M) \cap \overset{\circ}{B}(y,r)$ an elementary property of Hausdorff measures (see Falconer (1990, p. 28, 2.9)) implies

$$
\begin{aligned}
(12.2) \qquad \frac{1}{c'^D}\mathcal{H}^D(M \cap g^{-1}(\overset{\circ}{B}(y,r))) &\le \mathcal{H}^D(g(M) \cap \overset{\circ}{B}(y,r)) \\
&\le c'^D \mathcal{H}^D(M \cap g^{-1}(\overset{\circ}{B}(y,r))).
\end{aligned}
$$

Combining (12.1) and (12.2) yields

$$
\frac{1}{c}\left(\frac{1}{c'}\right)^{2D} r^D \le \mathcal{H}^D(g(M) \cap \overset{\circ}{B}(y,r)) \le cc'^{2D} r^D.
$$

Thus $\mathcal{H}^D_{|g(M)}$ has $g(M)$ as its support and $g(M)$ is regular of dimension D. □

We will now give some examples of regular sets of dimension D.

12.7 Example (Convex sets)

Let $K \subset \mathbb{R}^d$ be a nonempty compact convex set. The dimension D of K is defined as the dimension of the affine subspace of $\mathbb{R}^d$ spanned by K. We will show that K is a regular set of dimension D.

Without loss of generality we may assume that K spans $\mathbb{R}^d$ (otherwise we take the affine subspace generated by K and transform it by an affine isometry onto some $\mathbb{R}^d$). In this situation $D = d$, $\mathcal{H}^D_{|K}$ is just a non-zero multiple of the Lebesgue measure $\lambda^d_{|K}$, and, obviously,

$$
0 < \lambda^d(K) < \infty
$$

since $\operatorname{int} K \neq \emptyset$ (cf. Webster, 1994, p. 61, Theorem 2.3.1). By Remark 12.2 we may assume that $\mathbb{R}^d$ carries the l_2-norm. To prove that K is regular of dimension d it is, therefore, enough to show that $\lambda^d_{|K}$ is regular of dimension d. Let $r_0 > 0$ be arbitrary. The map $K \to \mathbb{R}$, $x \to \lambda^d(K \cap \overset{\circ}{B}(x, r_0))$ is continuous. Thus $c = \min_{x \in K} \lambda^d(K \cap \overset{\circ}{B}(x, r_0))$ exists. Since $\overset{\circ}{B}(x, r_0) \cap \operatorname{int} K \neq \emptyset$ we know that $c > 0$. For $0 < t \le 1$ and $x \in K$ the convexity of K yields

$$
x + t(\overset{\circ}{B}(0, r_0) \cap (K - x)) \subset K \cap \overset{\circ}{B}(x, tr_0).
$$

We deduce

$$\begin{aligned}\lambda^d(K \cap \mathring{B}(x, tr_0)) &\geq t^d \lambda^d(\mathring{B}(0, r_0) \cap (K - x)) \\ &= t^d \lambda^d(\mathring{B}(x, r_0) \cap K) \\ &\geq \frac{c}{r_0^d}(tr_0)^d.\end{aligned}$$

Thus we obtain, for $r \in (0, r_0)$,

$$\begin{aligned}\frac{c}{r_0^d} r^d \leq \lambda^d(K \cap \mathring{B}(x, r)) &\leq \lambda^d(\mathring{B}(x, r)) \\ &\leq \lambda^d(\mathring{B}(0, 1)) r^d.\end{aligned}$$

12.8 Example (Surfaces of convex sets)
Let $K \subset \mathbb{R}^d$ be nonempty, convex, and compact. Let $\operatorname{bd} K$ denote the relative boundary of K, i. e., the boundary of K relative to the affine subspace spanned by K. Let $D+1$ be the dimension of K. We will show that $\operatorname{bd} K$ is regular of dimension D.

As above we may assume that $D + 1 = d$ and that $\mathbb{R}^d$ carries the l_2-norm.
For $x, y \in \mathbb{R}^d$ the number $\langle x, y \rangle \in \mathbb{R}$ is the standard scalar product of x and y, i. e., if $x = (x_1, \ldots, x_d)$ and $y = (y_1, \ldots, y_d)$ then

$$\langle x, y \rangle = \sum_{i=1}^{d} x_i y_i. \tag{12.3}$$

Our claim is that, for every $x \in \partial K = \operatorname{bd} K$, there is an $r > 0$, a bi-Lipschitz map $f \colon B(x, r) \to \mathbb{R}^d$ and a hyperplane $H \subset \mathbb{R}^d$ with

$$f(B(x, r) \cap \partial K) = f(B(x, r)) \cap H.$$

Once this has been proved the argument is finished as follows. The set $f(\mathring{B}(x, r))$ is open. Hence there is an $s > 0$ with $B(f(x), s) \subset f(B(x, r))$. By Example 12.7 the set $B(f(x), s) \cap H$ is regular of dimension $d - 1$. The map $g = f^{-1}$ from $f(\mathring{B}(x, r))$ to $\mathring{B}(x, r)$ maps $B(f(x), s) \cap H$ onto $f^{-1}(B(f(x), s)) \cap \partial K$. Since g is bi-Lipschitz Lemma 12.6 implies that $f^{-1}(B(f(x), s) \cap \partial K$ is regular of dimension $d-1$. Moreover, $f^{-1}(B(f(x), s)) \cap \partial K$ is a neighbourhood of x in ∂K. By Lemma 12.5 ∂K is a regular set of dimension $d - 1$.

To prove the claim let $x \in \partial K$ be arbitrary. Then there exists a $\delta > 0$ and a unit vector $u \in \mathbb{R}^d$ with $x + \delta u \in \operatorname{int} K$. Moreover there is an $r > 0$ with $B(x + \delta u, r) \subset \operatorname{int} K$. Since $x \notin \operatorname{int} K$ we know that $r < \delta$. Define $\varphi \colon B(x, r) \to \mathbb{R}$ by $\varphi(y) = \min\{t \in \mathbb{R} \colon y + tu \in K\}$. Then φ is well defined since for an arbitrary $y \in B(x, r)$ we have $y + \delta u \in B(x + \delta u, r) \subset \operatorname{int} K$.

For $y \in \partial K$ the point $y + 0 \cdot u$ belongs to K so that $\varphi(y) \leq 0$. Since the open line segment between $y + \varphi(y)u$ and $y + \delta u$ lies in $\operatorname{int} K$ the fact that y belongs to ∂K

implies $\varphi(y) \geq 0$. Thus $y \in \partial K$ yields $\varphi(y) = 0$. Obviously every $y \in B(x,r)$ with $\varphi(y) = 0$ belongs to ∂K.

Next we will show that φ is convex. Let $y, y' \in B(x,r)$ and $t \in [0,1]$ be arbitrary. Then $ty + (1-t)y' + (t\varphi(y) + (1-t)\varphi(y'))u = t(y + \varphi(y)u) + (1-t)(y' + \varphi(y')u) \in K$, hence

$$\varphi(ty + (1-t)y') \leq t\varphi(y) + (1-t)\varphi(y').$$

A convex function on an open set is locally Lipschitz (see Webster, 1994, p. 224/225, proof of Theorem 5.51). Thus, by making r a bit smaller if necessary we may assume that φ is Lipschitz, i.e., there is a $c > 0$ with $|\varphi(y) - \varphi(y')| \leq c\|y - y'\|$ for all $y, y' \in B(x,r)$. Define $H = \{y \in \mathbb{R}^d \colon \langle y, u \rangle = 0\}$ and let P_H be the orthoghonal projection onto H. Define $f \colon B(x,r) \to R^d$ by $f(y) = P_H(y) - \varphi(y)u$. Then f is a Lipschitz map, since

$$\begin{aligned} \|f(y) - f(y')\| &\leq \|P_H(y) - P_H(y')\| + |\varphi(y) - \varphi(y')\| \\ &\leq \|y - y'\| + c\|y - y'\| \\ &\leq (1+c)\|y - y'\| \end{aligned} \tag{12.4}$$

Now we will show that there is a constant $c' > 0$ such that

$$\|f(y) - f(y')\| \geq c'\|y - y'\| \tag{12.5}$$

for all $y, y' \in B(x,r)$.
Let $y, y' \in B(x,r)$ be arbitrary, set $t = \langle x, u \rangle$ and define $z = P_H(y) + tu$, $z' = P_H(y') + tu$. Since u is orthogonal to H we deduce

$$\begin{aligned} \|z - x\|^2 &= \|P_H(z - x)\|^2 + \|tu - \langle x, u \rangle u\|^2 \\ &= \|P_H(z) - P_H(x)\|^2. \end{aligned}$$

Since $P_H(z) = P_H(y)$ we get

$$\|z - x\|^2 \leq \|y - x\|^2 \leq r^2,$$

hence $z \in B(x,r)$.
Similarly, $z' \in B(x,r)$.
By the definition of φ we have

$$\begin{aligned} \varphi(z) &= \min\{s \in \mathbb{R} \colon z + su \in K\} \\ &= \min\{s \in \mathbb{R} \colon P_H(y) + tu + su \in K\} \\ &= \min\{s \in \mathbb{R} \colon P_H(y) + \langle y, u \rangle u + (t + s - \langle y, u \rangle)u \in K\}. \end{aligned}$$

Since $y = P_H(y) + \langle y, u \rangle u$ we obtain

$$\begin{aligned} \varphi(z) &= \min\{s \in \mathbb{R} \colon y + (t + s - \langle y, u \rangle) \in K\} \\ &= \langle y, u \rangle - t + \varphi(y) \end{aligned} \tag{12.6}$$

For the same reason

$$\varphi(z') = \langle y', u\rangle - t + \varphi(y'). \tag{12.7}$$

Using (12.4), (12.6) and (12.7) and the definition of z and z' leads to

$$\begin{aligned} |\varphi(y) - \varphi(y')| &= |\varphi(z) + t - \langle y, u\rangle - \varphi(z') - t + \langle y', u\rangle| \\ &\geq |\langle y - y', u\rangle| - (1+c)\|z - z'\| \\ &= |\langle y - y', u\rangle| - (1+c)\|P_H(y) - P_H(y')\|. \end{aligned}$$

If $|\langle y - y', u\rangle| \geq 2(1+c)\|P_H(y) - P_H(y')\|$ then

$$|\varphi(y) - \varphi(y')| \geq \frac{1}{2}|\langle y - y', u\rangle|$$

hence

$$\begin{aligned} \|y - y'\|^2 &= \|P_H y - P_H y'\|^2 + |\langle y - y', u\rangle|^2 \\ &\leq 4\|P_H y - P_H y'\|^2 + 4|\varphi(y) - \varphi(y')|^2 \\ &= 4\|f(y) - f(y')\|^2 \end{aligned} \tag{12.8}$$

If $|\langle y - y', u\rangle| < 2(1+c)\|P_H u - P_H y'\|$ then

$$\begin{aligned} \|y - y'\|^2 &= \|P_H y - P_H y'\|^2 + |\langle y - y', u\rangle|^2 \\ &\leq (4(1+c)^2 + 1)\|P_H y - P_H y'\|^2 \\ &\leq 4((1+c)^2 + 1)\|f(y) - f(y')\|^2 \end{aligned} \tag{12.9}$$

Combining (12.8) and (12.9) yields (12.5), so that f is bi-Lipschitz.

Moreover we have

$$f(B(x,r) \cap \partial K)) = H \cap f(B(x,r)). \tag{12.10}$$

This can be seen as follows:
If $y \in \partial K \cap B(x,r)$ then $\varphi(y) = 0$ and, hence,

$$f(y) = P_H(y) \in H \cap f(B(x,r))$$

If $y \in B(x,r)$ and $f(y) \in H$ then $\varphi(y) = 0$, hence, $y \in \partial K \cap B(x,r)$.

12.9 Example (Compact differentiable manifolds)
Let $D \in \{1, \ldots, d-1\}$ and M be a compact D-dimensional C^1-submanifold of $\mathbb{R}^d$. Then M is regular of dimension D.

To prove this we will show that every point in M has a regular neighbourhood of dimension D in M. The claim then follows from Lemma 12.5. Let $x \in M$ be arbitrary. By Federer (1969, p. 231, 3.1.19) there exists an open neighbourhood U of x, an open set V in $\mathbb{R}^d$, a D-dimensional vector subspace W of $\mathbb{R}^d$, and a C^1-diffeomorphism $f\colon U \to V$ with $f(M \cap U) = W \cap V$. Let $r > 0$ be such that $B(f(x), r) \subset V$. By Example 12.7 the set $B(f(x), r) \cap W$ is regular of dimension D, $g = f^{-1}$ is bi-Lipschitz and maps $B(f(x), r) \cap W$ onto $f^{-1}(B(f(x), r)) \cap M$. By Lemma 12.6 $f^{-1}(B(f(x), r)) \cap M$ is regular of dimension D. Since $f^{-1}(B(f(x), r)) \cap M$ is obviously a neighbourhood of x in M the argument is finished.

12.10 Example (Self-similar sets)
Let $S_1, \ldots, S_N \colon \mathbb{R}^d \to \mathbb{R}^d$ be contracting similarity transformations with scaling numbers $s_1, \ldots, s_N \in (0,1)$, i.e., for $i \in \{1, \ldots, N\}$ and all $x, y \in \mathbb{R}^d$,

$$\|S_i x - S_i y\| = s_i \|x - y\|.$$

It was shown by Hutchinson (1981) that there is a unique non-empty compact set A in $\mathbb{R}^d$ with

$$A = S_1(A) \cup \ldots \cup S_N(A).$$

This set is called the self-similar set corresponding to $S_1, \ldots, S_N$. It is easy to see that there is a unique real number $D \geq 0$ with

$$\sum_{i=1}^{N} s_i^D = 1,$$

the similarity dimension of $(S_1, \ldots, S_N)$.
$(S_1, \ldots, S_N)$ is said to satisfy the open set condition (OSC) if there exists a nonempty open set $U \subset \mathbb{R}^d$ with $S_i(U) \subset U$ and $S_i(U) \cap S_j(U) = \emptyset$ for $i \neq j$, $i, j = 1, \ldots, N$.
Recovering an older result of Moran (1946), Hutchinson (1981, Theorem (1)) proved that

$$0 < \mathcal{H}^D(A) < \infty$$

if $(S_1, \ldots, S_N)$ satisfies the open set condition. Hutchinson (1981, p. 737/738) also showed that there are constants $c_1, c_2 > 0$ and $r_0 > 0$ with

$$c_1 r^D \leq \mathcal{H}^D(A \cap \mathring{B}(x, r)) \leq c_2 r^D$$

for all $x \in A$ and all $r \in (0, r_0)$.
In our language this means that A is regular of dimension D.
Classical examples of self-similar sets satisfying the OSC are the Cantor set, the Sierpinski gasket, and the von Koch curve. The **Cantor set** is the self-similar set corresponding to the two contractions $S_1, S_2 \colon I\!R \to I\!R$ with $S_1(x) = \frac{1}{3}x$ and $S_2(x) = \frac{1}{3}x + \frac{2}{3}$. The **Sierpinski gasket** corresponds to the three contractions $S_1, S_2, S_3 \colon I\!R^2 \to I\!R^2$ with $S_1 x = \frac{1}{2}x$, $S_2 x = \frac{1}{2}x + (\frac{1}{2}, 0)$ and $S_3 x = \frac{1}{2}x + (\frac{1}{3}, \frac{\sqrt{3}}{4})$. The **von Koch curve** is generated by the four contractions $S_1, S_2, S_3, S_4 \colon I\!R^2 \to I\!R^2$ defined by $S_1(x) = \frac{1}{3}x$, $S_2(x) = \frac{1}{3}D_{\frac{\pi}{3}}x + (\frac{1}{3}, 0)$ $S_3(x) = \frac{1}{3}D_{-\frac{\pi}{3}}x + (\frac{1}{2}, \frac{1}{12}\sqrt{3})$, $S_3(x) = \frac{1}{3}x + (\frac{2}{3}, 0)$, where D_φ is the counter-clockwise rotation about the origin with angle φ. By Hutchinson's result these sets are regular of dimension D, for $d = \frac{\log 2}{\log 3}$, $D = \frac{\log 3}{\log 2}$, and $D = \frac{\log 4}{\log 3}$ respectively. For more information about these and other self-similar sets we refer the reader to the book of Falconer (1990).

Before we deal with the quantization of regular measures of dimension D let us mention the following characterization of these measures.

12.11 Proposition
Let μ be a finite measure on $\mathbb{R}^d$ with compact support K. Then the following properties are equivalent:

(i) μ is regular of dimension D.

(ii) K is regular of dimension D and there exists a $c > 0$ with $\frac{1}{c}\mathcal{H}^D_{|K} \leq \mu \leq c\mathcal{H}^D_{|K}$.

Proof

That (i) implies (ii) is an immediate consequence of a proposition in Falconer(1990, Proposition 4.9) and the definition of regularity of dimension D.
The converse implication (ii) $\Rightarrow$ (i) follows from the definition of regularity of dimension D. □

12.2 Asymptotics for the quantization error

12.12 Proposition
Let P be a probability on $\mathbb{R}^d$. Assume that there is a $c > 0$ and an $r_0 > 0$ such that

$$P(\overset{\circ}{B}(a,r)) \leq cr^D \tag{12.11}$$

for all $a \in \operatorname{supp}(P)$ and all $r \in (0, r_0)$.
Then there exists a constant $c' > 0$ with

$$\int_B \|x-a\|\, dP(x) \geq c' P(B)^{1+\frac{1}{D}} \tag{12.12}$$

for all $a \in \mathbb{R}^d$ and all Borel sets $B \subset \mathbb{R}^d$.

Proof

By Lemma 12.3 there is a $\tilde{c} > 0$ with $P(\overset{\circ}{B}(a,r)) \leq \tilde{c}r^D$ for all $a \in \mathbb{R}^d$ and all $r > 0$. Let $a \in \mathbb{R}^d$ and let B be a Borel subset of $\mathbb{R}^d$. If $P(B) = 0$ then the conclusion (12.12) obviously holds. Let us assume $P(B) > 0$ and set

$$r_B = \inf\{r > 0\colon P(\overset{\circ}{B}(a,r)) \geq \tfrac{1}{2}P(B)\}.$$

Since $\lim_{r\to\infty} P(\overset{\circ}{B}(a,r)) = 1 \geq P(B)$ there is an $r > 0$ with $P(\overset{\circ}{B}(a,r)) \geq \frac{1}{2}P(B)$. Hence, $r_B < \infty$. For $r > r_B$ we have

$$\tilde{c}r^D \geq P(\overset{\circ}{B}(a,r)) \geq \tfrac{1}{2}P(B)$$

which implies

$$\tilde{c}r_B^D \geq \frac{1}{2}P(B). \tag{12.13}$$

For $r < r_B$ we have

$$P(\overset{\circ}{B}(a,r)) < \tfrac{1}{2}P(B)$$

Since $P(B) > 0$ there is an $r > 0$ with

$$P(\mathring{B}(a,r)) \leq \tilde{c}r^D < \tfrac{1}{2}P(B).$$

If $(r_n)_{n\in\mathbb{N}}$ is any increasing sequence with $r_n < r_B$ and $\lim_{n\to\alpha} r_n = r_B$ we deduce from $\mathring{B}(a,r_B) = \bigcup_{n\in\mathbb{N}} \mathring{B}(a,r_n)$ that

$$P(\mathring{B}(a,r_B)) = \lim_{n\to\infty} P(\mathring{B}(a,r_n)) \leq \frac{1}{2}P(B) \tag{12.14}$$

Using (12.14) and (12.13) we get

$$\begin{aligned}\int_B \|x-a\|\,dP(x) &\geq \int_{B\setminus\mathring{B}(a,r_B)} \|x-a\|\,dP(x)\\ &\geq r_B P(B\setminus\mathring{B}(a,r_B))\\ &\geq r_B(P(B) - P(\mathring{B}(a,r_B)))\\ &\geq \frac{1}{2}r_B P(B)\\ &\geq \frac{1}{2}(\frac{1}{2\tilde{c}})^{\frac{1}{D}}(P(B))^{1+\frac{1}{D}}.\end{aligned}$$

□

12.13 Corollary
Let P be a probability on $\mathbb{R}^d$. Then the following conditions are equivalent:

(i) There exists a $c > 0$ with

$$P(\mathring{B}(a,r)) \leq cr^D$$

for all $r > 0$ and all $a \in \mathbb{R}^d$.

(ii) There exists a $c' > 0$ with

$$\int_{\mathring{B}(a,r)} \|x-a\|\,dP(x) \geq c'P(\mathring{B}(a,r))^{1+\frac{1}{D}}$$

for all $r > 0$ and all $a \in \mathbb{R}^d$.

(iii) There exists a $c'' > 0$ with

$$\int_B \|x-a\|\,dP(x) \geq c''P(B)^{1+\frac{1}{D}}$$

for all $a \in \mathbb{R}^d$ and all Borel sets $B \subset \mathbb{R}^d$.

Proof

That (i) implies (iii) is Proposition 12.12 and that (iii) implies (ii) follows by setting $B = \mathring{B}(a,r)$. It remains to show that (ii) implies (i). Obviously, (ii) yields

$$rP(\mathring{B}(a,r)) \geq \int_{\mathring{B}(a,r)} \|x-a\|\, dP(x) \geq c' P(\mathring{B}(a,r))^{1+\frac{1}{D}}$$

and, hence,

$$(\tfrac{1}{c'})^D r^D \geq P(\mathring{B}(a,r)).$$

□

12.14 Corollary

Let P be a probability on $\mathbb{R}^d$. Suppose that there is a $c > 0$ and an $r_0 > 0$ with

$$P(\mathring{B}(a,r)) \leq cr^D$$

for all $a \in \operatorname{supp}(P)$ and all $r \in (0, r_0)$. Then there is a constant $b > 0$ such that, for every P-packing $\{B_1, \ldots, B_n\}$ in $\mathbb{R}^d$ with $P(\mathbb{R}^d \setminus \bigcup_{i=1}^n B_i) = 0$ and all $a_1, \ldots, a_n \in \mathbb{R}^d$:

$$n\Big(\sum_{i=1}^n \int_{B_i} \|x-a_i\|\, d\mu(x)\Big)^D \geq b.$$

Proof

Without loss of generality $D > 0$. Set $p = 1 + \frac{1}{D}$ and $q = 1 + D$. Then we have $\frac{1}{q} + \frac{1}{p} = 1$ and $p > 1$. Hölder's inequality yields

$$\begin{aligned} S &:= \sum_{i=1}^n 1 \Big(\int_{B_i} \|x-a_i\|\, dP(x)\Big)^{\frac{D}{1+D}} \\ &\leq \Big(\sum_{i=1}^n 1^q\Big)^{\frac{1}{q}} \Big(\sum_{i=1}^n \Big(\int_{B_i} \|x-a_i\|\, dP(x)\Big)^{\frac{D}{D+1}p}\Big)^{\frac{1}{p}} \\ &= n^{\frac{1}{q}} \Big(\sum_{i=1}^n \int_{B_i} \|x-a_i\|\, dP(x)\Big)^{\frac{1}{p}} \end{aligned}$$

This implies

$$S^q \leq n\Big(\sum_{i=1}^n \int_{B_i} \|x-a_i\|\, d\mu(x)\Big)^D.$$

Using Proposition 12.12 we get for the constant $c' > 0$ of that proposition

$$S = \sum_{i=1}^{n} \left(\int_{B_i} \|x - a_i\| \, dP(x) \right)^{\frac{D}{D+1}} \geq \sum_{i=1}^{n} (c')^{\frac{D}{D+1}} P(B_i) = (c')^{\frac{D}{D+1}}.$$

Thus, the corollary holds, if we set $b = (c')^D$. □

12.15 Proposition

Let P be a probability on $\mathbb{R}^d$. Suppose that there are constants $c > 0$ and $r_0 > 0$ with

$$P(\overset{\circ}{B}(a,r)) \leq cr^D$$

for every $a \in \operatorname{supp}(P)$ and every $r \in (0, r_0)$. Then

$$\liminf_{n \to \infty} n e_{n,1}(P)^D > 0.$$

Proof

Let $\alpha_n \in \mathcal{C}_{n,1}$ and let $\{A_a : a \in \alpha_n\}$ be a Voronoi partition of $\mathbb{R}^d$ with respect to α_n. By Corollary 12.14 we have

$$n e_{n,1}^D = n \left(\sum_{a \in \alpha_n} \int_{A_a} \|x - a\| \, dP(x) \right)^D \geq b > 0,$$

where b is as in Corollary 12.14. This implies the proposition. □

12.16 Corollary

Let P be a probability on $\mathbb{R}^d$. Then

$$\dim_H(P) \leq \underline{D}_r(P)$$

for all $r \geq 1$.

Proof

By Corollary 11.4 it suffices to prove $\dim_H(P) \leq \underline{D}_1(P)$. If $\dim_H(P) = 0$ then there is nothing to show. So let $\dim_H(P) > 0$ and let t with $0 < t < \dim_H(P)$ be arbitrary. By Falconer (1997, Prop. 10.3) we have

$$\dim_H(P) = \inf\{s \in \mathbb{R} : \liminf_{r \to 0} \frac{\log P(\overset{\circ}{B}(x,r))}{\log r} \leq s \text{ for } P-\text{a.e. } x\}.$$

This implies

$$P(\{x \in \mathbb{R}^d : \liminf_{r \to 0} \frac{\log P(\overset{\circ}{B}(x,r))}{\log r} > t\}) > 0$$

and, hence,

$$P(\{x \in \mathbb{R}^d : \exists r_x > 0 \ \forall r \le r_x : P(\overset{\circ}{B}(x,r)) \le r^t\}) > 0.$$

Thus, there exists a compact set $K \subset \mathbb{R}^d$ with $P(K) > 0$ and an $r_0 > 0$ such that

$$P(\overset{\circ}{B}(x,r)) \le r^t$$

for all $x \in K$ and all $r \le r_0$.
Set $Q = \frac{1}{P(K)} P_{|K}$. Then it follows that

$$Q(\overset{\circ}{B}(x,r)) \le \frac{1}{P(K)} r^t$$

for all $x \in \text{supp}(Q)$ and all $r \le r_0$ and Proposition 12.15 yields

$$\liminf_{n\to\infty} n e_{n,1}(Q)^t > 0.$$

Since $e_{n,1}(Q) \le \frac{1}{P(K)} e_{n,1}(P)$ this leads to

$$\liminf_{n\to\infty} n e_{n,1}(P)^t > 0.$$

Using Proposition 11.3 we deduce

$$t \le \underline{D}_1.$$

Since $t < \dim_H(P)$ was arbitrary this implies

$$\dim_H(P) \le \underline{D}_1.$$

□

12.17 Proposition
Let P be a probability on $\mathbb{R}^d$ with compact support K. Assume that there is a $c > 0$ and an $r_0 > 0$ with

$$cr^D \le P(\overset{\circ}{B}(a,r))$$

for every $a \in K$ and every $r \in (0, r_0)$.
Then

$$\limsup_{n\to\infty} n e_{n,\infty}(P)^D < \infty.$$

Proof

If there is an $n_0 \in \mathbb{N}$ with $e_{n_0,\infty} = 0$ then $e_{n,\infty} = 0$ for all $n \ge n_0$ and the assertion of the proposition is obvious. So let us assume that $e_{n,\infty} > 0$ for all $n \in \mathbb{N}$. Since K is compact there exists a finite set $\alpha_n \subset K$ of maximum cardinality satisfying $\|x - y\| \ge e_{n,\infty}$ for all $x, y \in \alpha_n$ with $x \ne y$.

We will show that $|\alpha_n| > n$. Assume the contrary. Then we know that

$$e_{n,\infty} \leq \sup_{x \in K} d_{\alpha_n}(x).$$

Hence there exists a $y \in K$ with $\|y - a\| \geq e_{n,\infty}$ for all $a \in \alpha_n$, which contradicts the maximality of α_n.

For $x, y \in \alpha_n$ with $x \neq y$ we have

$$\overset{\circ}{B}(x, \tfrac{1}{2}e_{n,\infty}) \cap \overset{\circ}{B}(y, \tfrac{1}{2}e_{n,\infty}) = \emptyset$$

hence

$$1 = P(K) \geq P\Big(\bigcup_{a \in \alpha_n} \overset{\circ}{B}(a, \tfrac{1}{2}e_{n,\infty})\Big) = \sum_{a \in \alpha_n} P\big(\overset{\circ}{B}(a, \tfrac{1}{2}e_{n,\infty})\big).$$

Due to $\lim_{n \to \infty} e_{n,\infty} = 0$ there is an $n_1 \in \mathbb{N}$ with

$$\tfrac{1}{2}\, e_{n,\infty} < r_0$$

for all $n \geq n_1$. Thus, for $n \geq n_1$,

$$\begin{aligned} 1 &\geq \sum_{a \in \alpha_n} c(\frac{1}{2}e_{n,\infty})^D \\ &= c|\alpha_n|(\frac{1}{2}e_{n,\infty})^D \geq \frac{c}{2^D} n\, e_{n,\infty}^D \end{aligned}$$

and, therefore,

$$n e_{n,\infty}^D \leq \frac{2^D}{c}.$$

This proves the proposition. □

Now we can formulate and prove the main result concerning the asymptotics of quantization errors for regular probabilities.

12.18 Theorem

Let P be a regular probability of dimension D on $\mathbb{R}^d$. Then, for $1 \leq r \leq \infty$,

$$0 < \liminf_{n \to \infty} n e_{n,r}(P)^D \leq \limsup_{n \to \infty} n e_{n,r}(P)^D < \infty. \tag{12.15}$$

In particular the quantization dimension $D_r(P)$ agrees with D which is also the Hausdorff dimension of the support of P.

Proof

The inequality (12.15) follows immediately from Propositions 12.15, 12.17 and Lemma 10.1 (a). The remaining statements follow from Corollary 11.4 and Proposition 12.11. □

12.19 Remark
It remains an open question for which regular probabilities P of dimension D the limit

$$\lim_{n\to\infty} ne_{n,r}^D$$

exists in $(0,\infty)$. Recall from (6.4) that in this situation, for $1 \leq r < \infty$,

$$Q_r(P) = \lim_{n\to\infty} n^{\frac{r}{D}} V_{n,r}(P) = \big(\lim_{n\to\infty} ne_{n,r}^D(P)\big)^{\frac{r}{D}}$$

is called the r-th quantization coefficient of P. It follows from Theorem 6.2 that for the normalized volume measure P of a convex compact set the r-th quantization coefficient exists. We conjecture that the same is true for the normalized surface measures on convex compact sets and compact C^1-manifolds. For the natural Hausdorff measure on a self-similar set the quantization coefficients exist in some cases while in other cases (like the classical Cantor set) they need not exist. We will discuss measures on self-similar sets in Section 14.

Notes

The concept of regularity for sets and measures can be found in several books on geometric measure theory (see, for instance, David and Semmes (1993, 1997) and Mattila (1995)). Since there are several different notions of regularity for sets and measures the above regularity is sometimes called Ahlfors-David regularity (see Mattila, 1995, p. 92). The elementary results on regular sets of dimension D (Lemma 12.4–12.6) and the results concerning the regularity of convex sets and their boundaries as well as that of compact C^1-manifolds are probably well-known. We just could not find an explicit reference. To our knowledge the results concerning the quantization of regular sets and measures of dimension D as stated above are new. A good introduction to the theory of convex sets is Webster (1994). The basic theory concerning self-similar sets as well as many examples can be found in Barnsley (1988). For the canonical normalized Hausdorff measure P on a self-similar set with OSC the inequalities in (12.15) were first proved in Graf and Luschgy (1996). After this book had essentially been finished Pötzelberger (1998a) gave different conditions for a probability P to ensure that $0 < \liminf_{n\to\infty} ne_{n,2}(P)^D$ or $\limsup_{n\to\infty} ne_{n,2}(P)^D < \infty$ or $\lim_{n\to\infty} ne_{n,2}(P)^D$ exists (for the l_2-norm), where D is suitably chosen.

13 Rectifiable curves

Here we consider the length measures on rectifiable curves. These measures can be obtained by restricting the one-dimensional Hausdorff measure to the given rectifiable curve. In this way we get an elementary class of singular measures of quantization dimension 1 for which the quantization coefficients exist and will be calculated. Nevertheless there are simple examples that show that the length measure on a rectifiable curve need not be regular of dimension 1 (see below).

In this section $\| \ \|$ will always denote an euclidean norm on $\mathbb{R}^d$. First we will collect some basic results about rectifiable curves.

13.1 Definition
Let $a, b \in \mathbb{R}$ with $a < b$. A **curve** (more exactly, a Jordan curve) Γ is the image of a continuous injection $\gamma\colon [a, b] \to \mathbb{R}^d$. γ is called a **parametrization** of Γ. A curve is called **rectifiable** if

$$L = L(\Gamma) = \sup\left\{\sum_{i=1}^{n} \|\gamma(t_i) - \gamma(t_{i-1})\|\colon n \in \mathbb{N},\ a = t_0 < \ldots < t_n = b\right\} < \infty.$$

L is called the **length of the curve** Γ.

13.2 Lemma
If $\Gamma \subset \mathbb{R}^d$ is a curve then $\mathcal{H}^1(\Gamma) = L(\Gamma)$, in particular $L(\Gamma)$ does not depend on the parametrization γ.

Proof

See Falconer (1985, p. 29, Lemma 3.2). □

13.3 Definition
Let Γ be a rectifiable curve of length L. A continuous injection $\gamma\colon [0, L] \to \Gamma$ is called a **parametrization by arc length** if $L(\gamma([0, t])) = t$ for all $t \in [0, L]$.

13.4 Remark
Every rectifiable curve admits a parametrization by arc length (see Falconer, 1985, p. 29).

13.5 Lemma
Let Γ be a rectifiable curve of length L and $\gamma\colon [0, L] \to \Gamma$ a parametrization by arc length. Let μ be 1-dimensional Lebesgue measure restricted to $[0, L]$. Then

(13.1) $\|\gamma(t) - \gamma(s)\| \leq |t - s|$ *for all* $s, t \in [0, L]$ *and* $\mathcal{H}^1_{|\Gamma} = \mu \circ \gamma^{-1}$.

Proof

The lemma follows immediately from Falconer (1985, p. 29, (3.2) and p. 30, Corollary 3.3). □

13.6 Lemma

Let Γ *be a rectifiable curve,* $x \in \Gamma$ *and* $r \in (0, \frac{1}{2}\operatorname{diam}\Gamma)$. *Then*

$$\mathcal{H}^1(\Gamma \cap \mathring{B}(x,r)) \geq r. \tag{13.2}$$

Proof

This follows from Falconer (1985, p. 30, Lemma 3.4). □

Inequality (13.2) is one half of the condition for regularity of dimension 1 of a rectifiable curve. That, in general, a rectifiable curve need not be regular of dimension 1 is shown by the following example.

13.7 Example

Let $m \in \mathbb{N}$, $m \geq 1$. Then there exist a continuous injection $\gamma_m \colon \left[\frac{1}{m+1}, \frac{1}{m}\right] \to \mathbb{R}^2$ with $\gamma_m\left(\frac{1}{m+1}\right) = \left(\frac{1}{m+1}, 0\right)$, $\gamma_m\left(\frac{1}{m}\right) = \left(\frac{1}{m}, 0\right)$, $\frac{1}{m+1} \leq \|\gamma_m(t)\| < \frac{1}{m}$ for all $t \in \left[\frac{1}{m+1}, \frac{1}{m}\right)$, and

$$l_m := L\left(\gamma_m\left(\left[\frac{1}{m+1}, \frac{1}{m}\right]\right)\right) = \max\left\{\frac{1}{m} - \frac{1}{m+1}, \frac{1}{\sqrt{m}} - \frac{1}{\sqrt{m+1}}\right\}.$$

Define $\gamma \colon [0,1] \to \mathbb{R}^2$ by

$$\gamma(t) = \begin{cases} 0, & t = 0 \\ \gamma_m(t), & \text{if } m \in \mathbb{N} \text{ with } t \in \left[\frac{1}{m+1}, \frac{1}{m}\right] \end{cases}$$

Then $\Gamma = \gamma([0,1])$ is a rectifiable curve since

$$L(\Gamma) \leq \sum_{m=1}^{\infty} l_m < \infty.$$

Moreover, for $m \geq 2$,

$$\mathcal{H}^1(\mathring{B}(0, \frac{1}{m}) \cap \Gamma) = \sum_{k=m}^{\infty} l_k$$

Since there exists an $m_0 \in \mathbb{N}$ with

$$\frac{1}{\sqrt{m}} - \frac{1}{\sqrt{m+1}} \geq \frac{1}{m} - \frac{1}{m+1}$$

for $m \geq m_0$, we have, for these m,

$$\sum_{k=m}^{\infty} l_k = \frac{1}{\sqrt{m}}$$

so that

$$\mathcal{H}^1(\mathring{B}(0, \frac{1}{m}) \cap \Gamma) = \frac{1}{\sqrt{m}}$$

This shows that

$$\sup_{r>0} \frac{1}{r} \mathcal{H}^1(\overset{\circ}{B}(0,r) \cap \Gamma) = +\infty$$

and Γ is not regular of dimension 1.

Before we come to the quantization of recitifiable curves we will prove a result concerning the distance of the n-optimal set of centers of order r from the support of the probability in question.

13.8 Lemma
Let P be a probability on $\mathbb{R}^d$ with compact support K. Let $1 \leq r < \infty$ and let α_n be an n-optimal set of centers for P of order r. Define

$$\delta_n = \max_{a \in \alpha_n} \max_{x \in W(a|\alpha_n) \cap K} \|x - a\| = \|d_{\alpha_n}\|_\infty .$$

Then

$$\frac{\delta_n}{2} \min_{x \in K} P(\overset{\circ}{B}(x, \frac{\delta_n}{2}))^{\frac{1}{r}} \leq e_{n,r}(P). \tag{13.3}$$

Proof

Let $a \in \alpha_n$ satisfy $\delta_n = \max_{x \in W(a|\alpha_n) \cap K} \|x - a\|$. Then there exists an $x \in W(a|\alpha_n) \cap K$ with $\delta_n = \|x - a\|$. For every $b \in \alpha_n$ we have

$$\|x - b\| \geq \|x - a\|.$$

For every $y \in \overset{\circ}{B}(x, \frac{1}{2}\delta_n)$ and every $b \in \alpha_n$ this yields

$$\|y - b\| \geq \|x - b\| - \|x - y\| \geq \|x - a\| - \|x - y\| = \delta_n - \|x - y\| \geq \frac{1}{2}\delta_n.$$

Using this inequality we deduce

$$\begin{aligned} e_{n,r}^r = V_{n,r} &= \int d_{\alpha_n}(z)^r \, dP(z) \\ &\geq \int_{K \cap \overset{\circ}{B}(x, \frac{1}{2}\delta_n)} d_{\alpha_n}(z)^r \, dP(z) \\ &\geq (\frac{1}{2}\delta_n)^r P(K \cap \overset{\circ}{B}(x, \frac{1}{2}\delta_n)) \end{aligned}$$

and the lemma is proved. □

13.9 Corollary
Let P be a probability on $\mathbb{R}^d$ with compact support K. Let $n \leq |K|$, $1 \leq r < \infty$ and let α_n be an n-optimal set of centers for P of order r. Then, for every $a \in \alpha_n$,

$$\frac{1}{2} d(a, K) \min_{y \in K} P(\overset{\circ}{B}(y, \frac{1}{2} d(a, K)))^{\frac{1}{r}} \leq e_{n,r}(P). \tag{13.4}$$

Proof

The corollary follows from Lemma 13.8 if one observes that

$$\delta_n \geq d(a, K)$$

for all $a \in \alpha_n$, since $W(a|\alpha_n) \cap K \neq \emptyset$ for all $a \in \alpha_n$ by (4.1). □

First we give a quantization result for line segments. For $x, y \in \mathbb{R}^d$ let $[x, y]$ be the line segment from x to y, i.e.

$$[x, y] = \{(1-t)x + ty \colon t \in [0, 1]\}.$$

It is a well-known fact that for $f\colon [0,1] \to \mathbb{R}^d$ with $f(t) = (1-t)x + ty$ we have

$$\frac{1}{\|x - y\|}\mathcal{H}^1_{|[x,y]} = U([0,1])^f. \tag{13.5}$$

(see Lemma (13.5)).

13.10 Lemma

Let $x, y \in \mathbb{R}^d$ with $x \neq y$ be given. Let $P = \frac{1}{\|x-y\|}\mathcal{H}^1_{|[x,y]}$. Then, for $1 \leq r < \infty$ and $n \geq 1$,

$$e_{n,r}(P) = \left(\frac{1}{1+r}\right)^{\frac{1}{r}} \frac{\|x - y\|}{2n}. \tag{13.6}$$

Proof

By the remark preceding the lemma we have $P = U([0,1])^f = U([0,1]) \circ f^{-1}$. Let $1 \leq r < \infty$.

"$\leq$": For $\alpha \in C_{n,r}(U([0,1])$ the set $\beta = f(\alpha)$ has n points. Hence we deduce

$$\begin{aligned} V_{n,r}(P) &\leq \int \min_{b\in\beta} \|x - b\|^r \, d(P(x) \\ &= \int \min_{a\in\alpha} \|f(t) - f(a)\|^r \, dU([0,1])(t) \end{aligned}$$

Since $\|f(t) - f(a)\| = |t - a| \; \|x - y\|$ we obtain

$$\begin{aligned} V_{n,r}(P) &\leq \|x - y\|^r \int \min_{a\in\alpha} |t - a|^r \, dU([0,1])(t) \\ &= \|x - y\|^r V_{n,r}(U([0,1])) \end{aligned}$$

According to Example 5.5 we have

$$V_{n,r}(U([0,1])) = \frac{1}{(1+r)(2n)^r}$$

This yields

$$e_{n,r}(P) = V_{n,r}(P)^{1/r} \leq \left(\frac{1}{1+r}\right) \frac{\|x - y\|}{2n}$$

"$\geq$": Let $\beta \in C_{n,r}(P)$. Since $\text{supp}(P) = [x,y]$ is convex and since the underlying norm is euclidean Remark 4.6 yields $\beta \subset [x,y]$. Let $\alpha \subset [0,1]$ equal $f^{-1}(\beta)$. Then α has n points and we get

$$\begin{aligned} V_{n,r}(P) &= \int \min_{b\in\beta} \|x-b\|^r \, d(P(x)) \\ &= \int \min_{a\in\alpha} \|f(t)-f(a)\|^r \, dU([0,1])(t) \\ &= \|x-y\|^r \int |t-a|^r \, dU([0,1])(t) \\ &\geq \|x-y\|^r V_{n,r}(U([0,1])) \\ &= \|x-y\|^r \frac{1}{(1+r)(2n)^r} \end{aligned}$$

and, hence,

$$e_{n,r}(P) = V_{n,r}(P)^{1/r} \geq \left(\frac{1}{1+r}\right)^{1/r} \frac{\|x-y\|}{2n}.$$

□

Now we will give a first lower bound for the quantization errors for a normalized one-dimensional Hausdorff measure on a rectifiable curve.

13.11 Lemma

Let $\gamma\colon [a,b] \to \mathbb{R}^d$ be a continuous injection which is a parametrization of the rectifiable curve Γ with length $L > 0$. Let $P = \frac{1}{L}\mathcal{H}^1_{|\Gamma}$. Then, for $1 \leq r < \infty$,

$$e_{n,r}(P) \geq \left(\frac{1}{(r+1)L}\right)^{\frac{1}{r}} \|\gamma(b)-\gamma(a)\|^{1+\frac{1}{r}} \frac{1}{2n}. \tag{13.7}$$

Proof

Let $G = \{(1-t)\gamma(a) + t\gamma(b)\colon t \in \mathbb{R}\}$ be the line through $\gamma(a)$ and $\gamma(b)$. Let P_G be the orthogonal projection onto G. By $[\gamma(a),\gamma(b)] = \{(1-t\gamma(a)+t\gamma(b)\colon t \in [0,1]\}$ we denote the line segment from $\gamma(a)$ to $\gamma(b)$. Let Q denote the image of P with respect to P_G. First we will show that

$$\mathcal{H}^1_{|P_G(\Gamma)} \leq \mathcal{H}^1_{|\Gamma} \circ P_G^{-1}. \tag{13.8}$$

Let B be a Borel set in $\mathbb{R}^d$. Then using the fact that

$$\|P_G(x) - P_G(y)\| \leq \|x-y\|$$

for all $x, y \in \mathbb{R}^d$ and Falconer (1985, p. 27, Proposition 2.2) we get

$$\begin{aligned} \mathcal{H}^1_{|\Gamma}(P_G^{-1}(B)) &= \mathcal{H}^1(P_G^{-1}(B) \cap \Gamma) \\ &\geq \mathcal{H}^1(P_G(P_G^{-1}(B) \cap \Gamma)) \\ &= \mathcal{H}^1(B \cap P_G(\Gamma)) \\ &= \mathcal{H}^1_{|P_G(\Gamma)}(B). \end{aligned}$$

Thus, (13.8) is proved.
Now let $\alpha \in C_{n,r}(P)$. Using (13.8), the fact that $[\gamma(a), \gamma(b)] \subset P_G(\Gamma)$, and Lemma 13.10 we obtain

$$\begin{aligned}
e_{n,r}^r = V_{n,r} &= \int_\Gamma d(x,\alpha)^r \, dP(x) \\
&= \frac{1}{L} \int_\Gamma d(x,\alpha)^r \, d\mathcal{H}^1_{|\Gamma}(x) \\
&\geq \frac{1}{L} \int_\Gamma d(P_G(x), P_G(\alpha))^r \, d\mathcal{H}^1_{|\Gamma}(x) \\
&= \frac{1}{L} \int_{P_G(\Gamma)} d(y, P_G(\alpha))^r \, d\mathcal{H}^1_{|\Gamma} \circ P_G^{-1}(y) \\
&\geq \frac{1}{L} \int_{P_G(\Gamma)} d(y, P_G(\alpha))^r \, d\mathcal{H}^1_{|P_G(\Gamma)}(y) \\
&\geq \frac{1}{L} \int_{[\gamma(a),\gamma(b)]} d(y, P_G(\alpha))^r \, d\mathcal{H}^1(y) \\
&\geq \frac{1}{L} \|\gamma(b) - \gamma(a)\| V_{n,r} \left(\frac{1}{\|\gamma(b) - \gamma(a)\|} \mathcal{H}_{|[\gamma(a),\gamma(b)]} \right) \\
&= \frac{1}{L} \frac{\|\gamma(b) - \gamma(a)\|}{1+r} \left(\frac{\|\gamma(b) - \gamma(a)\|}{2n} \right)^r
\end{aligned}$$

Thus, the lemma is proved. □

13.12 Theorem
Let $\Gamma \subset \mathbb{R}^d$ be a rectifiable curve, with length $L > 0$. Set $P = \frac{1}{L}\mathcal{H}^1_{|\Gamma}$. Then

(i) for $1 \leq r < \infty$, $\lim_{n\to\infty} n e_{n,r}(P) = Q_r([0,1])^{1/r} \mathcal{H}^1(\Gamma) = \left(\frac{1}{1+r}\right)^{\frac{1}{r}} \frac{L}{2}$,

(ii) $\lim_{n\to\infty} n e_{n,\infty}(P) = Q_\infty([0,1]) \mathcal{H}^1(\Gamma) = \frac{L}{2}$.

Proof

Let $\gamma\colon [0,L] \to \Gamma$ be a parametrization of Γ by arc length.
First we will show (i).

"$\geq$" : Let $0 = t_0 < t_1 < \ldots < t_m = L$ and choose $t_0^- = t_0 = t_0^+$, $t_m^- = t_m = t_m^+$ and, for $i \in \{1, \ldots, m-1\}$,

$$t_{i-1}^+ < t_i^- < t_i < t_i^+ < t_{i+1}^-.$$

Let $\Gamma_i = \gamma([t_{i-1}^+, t_i^-])$. Then we know that $\Gamma_i \cap \Gamma_j = \emptyset$ for $i \neq j$. Let $\delta = \min\{d(\Gamma_i, \Gamma_j) \colon i \neq j\}$, where $d(B, C) = \inf\{d(b, C) \colon b \in C\}$ for $B, C \subset \mathbb{R}^d$. Then we have $\delta > 0$ and $\delta \leq \operatorname{diam}(\Gamma)$. By Lemma 13.6 this implies

$$\mathcal{H}^1\left(\Gamma \cap \overset{\circ}{B}\left(x, \frac{\delta}{4}\right)\right) \geq \frac{\delta}{4}$$

for all $x \in \Gamma$, and hence,

$$\frac{\delta}{4} \min_{x \in \Gamma} P\left(\Gamma \cap \overset{\circ}{B}\left(x, \frac{\delta}{4}\right)\right)^{\frac{1}{r}} \geq \left(\frac{1}{L}\right)^{\frac{1}{r}} \left(\frac{\delta}{4}\right)^{1+\frac{1}{r}} \tag{13.9}$$

Let α_n be an n-optimal set of centers of order r. For $i = 1, \ldots, n$ set

$$\alpha_{n,i} = \{a \in \alpha_n \colon W(a|\alpha_n) \cap \Gamma_i \neq \emptyset\}.$$

We will show that there is an $n_0 \in \mathbb{N}$ such that

$$\alpha_{n,i} \cap \alpha_{n,j} = \emptyset \tag{13.10}$$

for $i \neq j$ and all $n \geq n_0$. Since $\lim_{n \to \infty} e_{n,r} = 0$ there exists an $n_0 \in \mathbb{N}$ with

$$e_{n_0,r} < \left(\frac{1}{L}\right)^{\frac{1}{r}} \left(\frac{1}{4}\delta\right)^{\frac{1+r}{r}}. \tag{13.11}$$

Using Lemma 13.8 we have

$$\frac{1}{2} \|d_{\alpha_n}\|_\infty \min_{y \in \Gamma} P\left(\overset{\circ}{B}\left(y, \frac{1}{2}\|d_{\alpha_n}\|_\infty\right)\right)^{\frac{1}{r}} \leq e_{n,r} \leq e_{n_0,r} \tag{13.12}$$

for all $n \geq n_0$. Observing that $t \to t \min_{y \in \Gamma} P(B(y,t))^{\frac{1}{r}}$ is non-decreasing and using (13.12), (13.11), and (13.9) we deduce

$$\|d_{\alpha_n}\|_\infty < \frac{\delta}{2}$$

for all $n \geq n_0$. By the definition of δ and $\alpha_{n,i}$ this yields (13.10). It follows from (13.10) that, for $n \geq n_0$,

$$\begin{aligned} n^r V_{n,r} &= n^r \int_\Gamma d(x, \alpha_n)^r \, dP(x) \\ &\geq n^r \sum_{i=1}^m \int_{\Gamma_i} d(x, \alpha_n)^r \, dP(x) \\ &= n^r \sum_{i=1}^m \int_{\Gamma_i} d(x, \alpha_{n,i})^r \, dP(x) \end{aligned}$$

Setting $n_i = |\alpha_{n,i}|$, $L_i = \mathcal{H}^1(\Gamma_i)$, and $P_i = \frac{1}{L_i}\mathcal{H}^1_{|\Gamma_i}$ we get

$$\begin{aligned} n^r V_{n,r} &\geq n^r \sum_{i=1}^m \frac{L_i}{L}\frac{1}{L_i}\int\limits_{\Gamma_i} d(x,\alpha_{n,i})^r \, d\mathcal{H}^1_{|\Gamma_i}(x) \\ &= n^r \sum_{i=1}^m \frac{L_i}{L} V_{n_i,r}(P_i) \end{aligned}$$

By Lemma 13.11 this leads to

$$\begin{aligned} n^r V_{n,r} &\geq n^r \sum_{i=1}^m \frac{L_i}{L}\frac{1}{(1+r)L_i}\|\gamma(t_{i-1}^+) - \gamma(t_i^-)\|^{1+r}\left(\frac{1}{2n_i}\right)^r \\ &\geq \frac{1}{2^r(1+r)L}\sum_{i=1}^m \|\gamma(t_{i-1}^+) - \gamma(t_i^-)\|^{1+r}\left(\frac{n_i}{n}\right)^{-r} \end{aligned} \tag{13.13}$$

Set $s_i = \|\gamma(t_{i-1}^+) - \gamma(t_i^-)\|^{1+r}$. It follows from Lemma 6.8 that

$$\sum_{i=1}^m s_i\left(\frac{n_i}{n}\right)^{-r} \geq \left(\sum_{i=1}^m s_i^{\frac{1}{1+r}}\right)^{1+r}, \tag{13.14}$$

hence

$$(ne_{n,r})^r = n^r V_{n,r} \geq \frac{1}{2^r(1+r)L}\left(\sum_{i=1}^m \|\gamma(t_{i-1}^+) - \gamma(t_i^-)\|\right)^{1+r}$$

For $t_i^- \to t_i$, $t_i^+ \to t_i$ we deduce

$$\liminf_{n\to\infty} ne_{n,r} \geq \frac{1}{2}\left(\frac{1}{(1+r)}\right)^{\frac{1}{r}}\left(\frac{1}{L}\right)^{\frac{1}{r}}\left(\sum_{i=1}^m \|\gamma(t_{i-1}) - \gamma(t_i)\|\right)^{\frac{1+r}{r}}$$

Since $L = \sup\{\sum_{i=1}^m \|\gamma(t_{i-1}) - \gamma(t_i)\| : a = t_0 < \ldots < t_m = b\}$ we obtain

$$\liminf_{n\to\infty} ne_{n,r} \geq \frac{1}{2}\left(\frac{1}{1+r}\right)^{\frac{1}{r}} L \tag{13.15}$$

"$\leq$" : Now let $\beta \subset [0,L]$ be of cardinality less than or equal to n and set $\alpha = \gamma(\beta)$. Using Lemma 13.5 we deduce

$$\begin{aligned} n^r V_{n,r} &\leq n^r \int \min_{a\in\alpha} \|x-a\|^r \, dP(x) \\ &= n^r \frac{1}{L}\int\limits_{[0,L]} \min_{b\in\beta} \|\gamma(t) - \gamma(b)\|^r \, dt \\ &\leq n^r \frac{1}{L}\int\limits_{[0,L]} \min_{b\in\beta} |t-b|^r \, dt \end{aligned} \tag{13.16}$$

Since β with $|\beta| \leq n$ was arbitrary in $[0, L]$ we deduce

$$n^r V_{n,r} \leq n^r V_{n,r}(U([0, L])$$

By Example 5.5 we have

$$V_{n,r}(U([0, L]) = \frac{L^r}{n^r(1+r)2^r}.$$

Thus we get

$$\limsup_{n\to\infty} n e_{n,r} \leq \left(\frac{1}{1+r}\right)^{\frac{1}{r}} \frac{L}{2} \tag{13.17}$$

Combining (13.15) and (13.17) yields the first part of the theorem.

Now we will prove part (ii).

"$\leq$" : Let α be as above.
Then using Lemma 13.5 we have

$$\begin{aligned} n e_{n,\infty}(P) &\leq n \max_{x\in\Gamma} \min_{a\in\alpha} \|x - a\| \\ &= n \max_{t\in[0,L]} \min_{b\in\beta} \|\gamma(t) - \gamma(b)\| \\ &\leq n \max_{t\in[0,L]} \min_{b\in\beta} |t - b|. \end{aligned}$$

Since β with $|\beta| \leq n$ was arbitrary in $[0, L]$ we get

$$n e_{n,\infty}(P) \leq n e_{n,\infty}(U([0, L]) = \frac{L}{2}.$$

Hence

$$\limsup_{n\to\infty} n e_{n,\infty}(P) \leq \frac{L}{2} \tag{13.18}$$

"$\geq$" : On the other hand (11.1) yields

$$L = \mathcal{H}^1(\Gamma) \leq 2 \liminf_{n\to\infty} n e_{n,\infty}(P)$$

so that

$$\liminf_{n\to\infty} n e_{n,\infty}(P) \geq \frac{L}{2}. \tag{13.19}$$

Combining (13.18) and (13.19) implies part (ii) of the theorem. □

13.13 Remark
In geometric measure theory a measure μ on $\mathbb{R}^d$ is called m-recitifiable if $m \in \{1, \dots, d\}$, μ is absolutely continuous with respect to $\mathcal{H}^m$, and μ is supported on a countable union of m-dimensional C^1-manifolds. A length measure on a rectifiable curve is 1-rectifiable in this sense.
We conjecture that, for all m-rectifiable measures μ on $\mathbb{R}^d$ and all $1 \leq r \leq \infty$,

$$\lim_{n \to \infty} n e_{n,r}(\mu)^m$$

exists in $(0, +\infty)$, i. e., every m-rectifiable measure has a r-th quantization coefficient.

Notes

The basic notions about rectifiable curves can, for instance, be found in the book of Falconer (1985, Chapter 3). A good introduction to the theory of rectifiable measures is given by Mattila (1995, §15–20).

14 Self-similar sets and measures

The class of self-similar measures has been a central object of studies in fractal geometry during the last two decades. Most self-similar measures are singular with respect to Lebesgue measure, but the restriction of Lebesgue measure to the d-dimensional cube is, for instance, also self-similar. In this section we determine the quantization dimension of all self-similar measures that satisfy a certain separation condition. As it turns out many self-similar measures have the property that their quantization dimension of order r is strictly increasing with r. In this respect they are different from all measures considered so far in this volume.

14.1 Basic notion and facts

Let $\| \; \|$ denote a norm on $\mathbb{R}^d$. A **contracting similarity transformation** S on $\mathbb{R}^d$ is a map $S\colon \mathbb{R}^d \to \mathbb{R}^d$ such that there is a constant $s \in (0,1)$ with

$$\|S(x) - S(y)\| = s\|x - y\| \tag{14.1}$$

for all $x, y \in \mathbb{R}^d$. s is called the **scaling number** or **contraction number** of S.
In what follows N is always a natural number greater than 1, $(S_1, \ldots, S_N)$ is an N-tuple of contracting similarity transformations of $\mathbb{R}^d$, and $(s_1, \ldots, s_N)$ is the corresponding N-tuple of scaling numbers.
Hutchinson (1981) has shown that there is always a unique nonempty compact subset A of $\mathbb{R}^d$ with

$$A = S_1(A) \cup \ldots \cup S_N(A). \tag{14.2}$$

A is called the **attractor** or **invariant set** of $(S_1, \ldots, S_N)$. Sometimes A is also called the self-similar set corresponding to $(S_1, \ldots, S_N)$.
It is easy to see that there exists a unique real number $D \geq 0$ with $\sum_{i=1}^{N} s_i^D = 1$, the **similarity dimension** of $(S_1, \ldots, S_N)$. Let $(p_1, \ldots, p_N)$ be a probability vector, i. e., $p_i \geq 0$ and $\sum_{i=1}^{N} p_i = 1$. Then Hutchinson (1981) showed that there is a unique probability measure P on $\mathbb{R}^d$ with

$$P = \sum_{i=1}^{N} p_i P \circ S_i^{-1}. \tag{14.3}$$

If $p_i > 0$ for all $i \in \{1, \ldots, N\}$ then the support of P equals the attractor A. P is called the **self-similar measure** corresponding to $(S_1, \ldots, S_N; p_1, \ldots, p_N)$. $(S_1, \ldots, S_N)$ satisfies the **strong separation condition** if

$$S_i(A) \cap S_j(A) = \emptyset$$

for $i \neq j$.
$(S_1, \dots, S_N)$ satisfies the **open set condition** (OSC) if there exists a nonempty open set $U \subset \mathbb{R}^d$ with $S_i(U) \subset U$ and $S_i(U) \cap S_j(U) = \emptyset$ for $i \neq j$.
Schief (1994) has shown that the open set U in the above definition can always be chosen to be bounded and satisfy $U \cap A \neq \emptyset$.
If $(S_1, \dots, S_N)$ satisfies the strong separation property than it also satisfies the open set condition.
If $(S_1, \dots, S_N)$ satisfies the OSC and P is the self-similar measure corresponding to $(S_1, \dots, S_N; s_1^D, \dots, s_N^D)$ then P is the normalized D-dimensional Hausdorff measure restricted to the attractor A, i. e. $0 < \mathcal{H}^D(A) < \infty$ and

$$P = \frac{1}{\mathcal{H}^D(A)} \mathcal{H}^D_{|A} \tag{14.4}$$

(see Hutchinson, 1981, p. 737/738).
By $\{1, \dots, N\}^*$ we denote the set of all words on the alphabet $1, \dots, N$ including the empty word $\emptyset$.
If $(q_1, \dots, q_N)$ is an N-tuple of real numbers and $\sigma = \sigma_1 \dots \sigma_n$ belongs to $\{1, \dots, N\}^*$ then define

$$q_\sigma = \begin{cases} 1, & \sigma = \emptyset \\ \prod_{i=1}^{n} q_{\sigma_i}, & \text{otherwise} \end{cases}$$

For a nonempty word $\sigma = \sigma_1 \dots \sigma_n \in \{1, \dots, N\}^*$ set

$$\sigma^- = \begin{cases} \emptyset, & n = 1 \\ \sigma_1 \dots \sigma_{n-1}, & n > 1. \end{cases}$$

If σ is a word then the **length of** σ, denoted by $|\sigma|$, is 0 if $\sigma = \emptyset$ and equal to n if $\sigma = \sigma_1 \dots \sigma_n$. For $m \leq |\sigma|$ let

$$\sigma_{|m} = \begin{cases} \emptyset, & m = 0 \\ \sigma_1 \dots \sigma_m, & m \geq 1 \end{cases}$$

be the restriction of σ to m. A natural order for words is defined by

$$\sigma \leq \tau \text{ iff } |\sigma| \leq |\tau| \text{ and } \tau_{||\sigma|} = \sigma.$$

For an infinite sequence $\eta \in \{1, \dots, N\}^{\mathbb{N}}$ the restriction $\eta_{|m}$ is defined in an analogous way. A word $\sigma \in \{1, \dots, N\}^*$ is a **predecessor** of η iff

$$\eta_{||\sigma|} = \sigma.$$

A finite set $\Gamma \subset \{1, \dots, N\}^*$ is called a **finite antichain** iff any two elements of Γ are incomparable with respect to the order given above. A finite antichain Γ is called maximal iff, for every finite antichain $\Gamma' \subset \{1, \dots, N\}^*$ with $\Gamma \subset \Gamma'$, we have $\Gamma = \Gamma'$.

A finite antichain Γ is maximal if and only if every sequence in $\{1,\ldots,N\}^{\mathbb{N}}$ has a precessor in Γ. If $(q_1,\ldots,q_N)$ is a probability vector and Γ is a maximal finite antichain then

(14.5)
$$\sum_{\sigma\in\Gamma} q_\sigma = 1.$$

If $0 < \varepsilon \le \min\{q_1,\ldots,q_N\}$ then

$$\Gamma(\varepsilon) = \{\sigma \in \{1,\ldots,N\}^*: q_{\sigma^-} \ge \varepsilon > q_\sigma\}$$

is a maximal finite antichain.
For $\sigma \in \{1,\ldots,N\}^*$ set

$$S_\sigma = \begin{cases} \mathrm{id}, & \sigma = \emptyset \\ S_{\sigma_1} \circ \ldots \circ S_{\sigma_n}, & \sigma = \sigma_1 \ldots \sigma_n, \end{cases}$$

where id is the identity on $\mathbb{R}^d$.

14.2 An upper bound for the quantization dimension

We use the notation introduced in 14.1. In the following $(p_1,\ldots,p_N)$ is always a probability vector with $p_i > 0$ for $i = 1,\ldots,N$ and P is the self-similar probability measure on $\mathbb{R}^d$ corresponding to $(S_1,\ldots,S_N; p_1,\ldots,p_N)$.

14.1 Lemma
For every $n \ge N$ and every $r \in [1,\infty)$,

(14.6)
$$V_{n,r}(P) \le \min\Big\{\sum_{i=1}^N p_i s_i^r V_{n_i,r}(P) : 1 \le n_i,\ \sum_{i=1}^N n_i \le n\Big\}.$$

Proof

Since S_i is a similarity transformation it follows immediately (see Lemma 3.2) that

(14.7)
$$V_{m,r}(P \circ S_i^{-1}) = s_i^r V_{m,r}(P)$$

for all $i \in \{1,\ldots,N\}$ and all $m \in \mathbb{N}$. Using (14.3) Lemma 4.14 (b) implies, for $n_1,\ldots,n_N \in \mathbb{N}$ with $n_i \ge 1$ and $\sum_{i=1}^N n_i \le n$,

$$V_{n,r}(P) \le \sum_{i=1}^N p_i V_{n_i,r}(P \circ S_i^{-1}).$$

Using (14.7) to substitute each summand on the right hand side yields the assertion of the lemma. □

14.2 Corollary
Let $\Gamma \subset \{1, \dots, N\}^*$ be a finite maximal antichain, $n \in \mathbb{N}$ with $n \geq |\Gamma|$ and $r \in [1, +\infty)$. Then

$$V_{n,r}(P) \leq \min\Big\{\sum_{\sigma \in \Gamma} p_\sigma s_\sigma^r V_{n_\sigma,r}(P) \colon 1 \leq n_\sigma, \sum_{\sigma \in \Gamma} n_\sigma \leq n\Big\} \tag{14.8}$$

Proof

The corollary follows from Lemma 14.1 by induction on $\max\{|\sigma| \colon \sigma \in \Gamma\}$. □

14.3 Lemma
For every $n \geq N$,

$$e_{n,\infty}(P) \leq \min\Big\{\max_{1 \leq i \leq N} s_i e_{n_i,\infty}(P) \colon 1 \leq n_i, \sum_{i=1}^{N} n_i \leq n\Big\}. \tag{14.9}$$

Proof

Since $\operatorname{supp}(P \circ S_i^{-1}) = S_i(A)$ and $A = \bigcup_{i=1}^{N} S_i(A)$ the lemma is an immediate consequence of Lemma 10.2(b) and Lemma 10.6(b). □

14.4 Lemma
Let $r \in [1 + \infty)$ be fixed. Then there exists exactly one number $\kappa_r \in (0, +\infty)$ with

$$\sum_{i=1}^{N} (p_i s_i^r)^{\frac{\kappa_r}{\kappa_r + r}} = 1. \tag{14.10}$$

Proof

Since $0 < p_i s_i^r < 1$ the function $t \to \sum_{i=1}^{N} (p_i s_i^r)^t$ is strictly decreasing and continuous. Since this function tends to N as t tends to 0 and takes a value less than 1 at $t = 1$ the intermediate value theorem implies the existence of a unique $t \in (0, 1)$ with $\sum_{i=1}^{N} (p_i s_i^r)^t = 1$. Then $\kappa_r = \frac{rt}{1-t}$ satisfies the conclusions of the lemma. □

14.5 Proposition
Let $r \in [1, +\infty)$ and let κ_r satisfy (14.10). Then

$$\limsup_{n \to \infty} n e_{n,r}^{\kappa_r} < \infty,$$

in particular, the upper quantization dimension $\overline{D}_r(P)$ of P is less than or equal to κ_r.

Proof

Let $q_i = (p_i s_i^r)^{\frac{\kappa_r}{r+\kappa_r}}$ and $\varepsilon_0 = \min\{q_1, \dots, q_N\}$. Then we have $\varepsilon_0 > 0$. Let $m, n \in \mathbb{N}$ be arbitrary with $\frac{m}{n} < \varepsilon_0^2$ and set $\varepsilon = \varepsilon_0^{-1}\frac{m}{n}$. For $\Gamma(\varepsilon) = \{\sigma \in \{1, \dots, N\}^* \colon q_{\sigma^-} \geq \varepsilon > q_\sigma\}$ it follows by (14.5) that

$$\begin{aligned} 1 &= \sum_{\sigma\in\Gamma(\varepsilon)} q_\sigma \\ &= \sum_{\sigma\in\Gamma(\varepsilon)} q_{\sigma^-} q_{\sigma_{|\sigma|}} \\ &\geq \varepsilon\varepsilon_0|\Gamma(\varepsilon)|, \end{aligned}$$

hence

$$|\Gamma(\varepsilon)| \leq (\varepsilon\varepsilon_0)^{-1} = \frac{n}{m}$$

Using Corollary 14.2 we deduce

$$\begin{aligned} V_{n,r}(P) &\leq \sum_{\sigma\in\Gamma(\varepsilon)} p_\sigma s_\sigma^r V_{m,r}(P) \\ &= \sum_{\sigma\in\Gamma(\varepsilon)} (p_\sigma s_\sigma^r)^{\frac{\kappa_r}{r+\kappa_r}} (p_\sigma s_\sigma^r)^{\frac{r}{r+\kappa_r}} V_{m,r}(P) \\ &\leq (\varepsilon^{\frac{r+\kappa_r}{\kappa_r}})^{\frac{r}{r+\kappa_r}} V_{m,r}(P) \sum_{\sigma\in\Gamma(\varepsilon)} q_\sigma \\ &= \varepsilon_0^{-\frac{r}{\kappa_r}} (\frac{m}{n})^{\frac{r}{\kappa_r}} V_{m,r}(P). \end{aligned}$$

Thus we obtain

$$n^{\frac{r}{\kappa_r}} V_{n,r}(P) \leq \varepsilon_0^{-\frac{r}{\kappa_r}} m^{\frac{r}{\kappa_r}} V_{m,r}(P)$$

which implies

$$n e_{n,r}(P)^{\kappa_r} \leq \varepsilon_0^{-1} m e_{m,r}(P)^{\kappa_r}.$$

Since, for fixed m, this inequality holds for all but a finite number of n we get

$$\begin{aligned} \limsup_{n\to\infty} n e_{n,r}(P)^{\kappa_r} &\leq \varepsilon_0^{-1} m e_{m,r}(P)^{\kappa_r} \\ &< +\infty \end{aligned}$$

and the proposition is proved. □

14.6 Remark

The proof of the preceding proposition shows that

$$(14.11)\quad \limsup_{n\to\infty} n e_{n,r}(P)^{\kappa_r} \leq \max\{(p_1 s_1^r)^{-\frac{\kappa_r}{r+\kappa_r}}, \dots, (p_N s_N^r)^{-\frac{\kappa_r}{r+\kappa_r}}\} \inf_{m\in\mathbb{N}} m e_{m,r}(P)^{\kappa_r}.$$

14.7 Proposition
Let D be the similarity dimension of $(S_1, \dots, S_N)$. Then

$$\limsup_{n\to\infty} n e_{n,\infty}(P)^D < +\infty,$$

in particular, the upper quantization dimension $\overline{D}_\infty(P)$ of P is less than or equal to D.

Proof

Set $q_i = s_i^D$ and $\varepsilon_0 = \min\{q_1, \dots, q_N\}$. Let $m, n \in \mathbb{N}$ be arbitrary with $\frac{m}{n} < \varepsilon_0^2$. Set $\varepsilon = \varepsilon_0^{-1}\frac{m}{n}$ and $\Gamma(\varepsilon) = \{\sigma \in \{1, \dots, N\}^* \colon q_{\sigma^-} \geq \varepsilon > q_\sigma\}$. Since $\Gamma(\varepsilon)$ is a maximal finite antichain it follows by Lemma 14.3 and an induction argument that

$$e_{n,\infty}(P)^D \leq \min\Big\{\max_{\sigma\in\Gamma(\varepsilon)} s_\sigma^D e_{n_\sigma,\infty}(P)^D \colon 1 \leq n_\sigma, \sum_{\sigma\in\Gamma(\varepsilon)} n_\sigma \leq n\Big\}.$$

As in the proof of Proposition 14.5 we see that $|\Gamma(\varepsilon)| \leq \frac{n}{m}$, so that

$$\begin{aligned} e_{n,\infty}(P)^D &\leq \max_{\sigma\in\Gamma(\varepsilon)} s_\sigma^D e_{m,\infty}(P)^D \\ &\leq \varepsilon e_{m,\infty}(P)^D \\ &= \varepsilon_0^{-1}\frac{m}{n} e_{m,\infty}(P)^D \end{aligned}$$

and, hence,

$$n e_{m,\infty}(P)^D \leq \varepsilon_0^{-1} m e_{m,\infty}(P)^D.$$

For fixed m this holds for all but finitely many n and yields

$$\limsup_{n\to\infty} n e_{n,\infty}(P)^D < +\infty.$$

The remaining statement of the proposition follows from Proposition 11.3 □

14.8 Remark
The proof of the preceding proposition shows that

$$\limsup_{n\to\infty} n e_{n,\infty}(P)^D \leq \max(s_1^{-D}, \dots, s_N^{-D}) \inf_{m\in\mathbb{N}} m e_{m,\infty}(P)^D. \tag{14.12}$$

14.3 A lower bound for the quantization dimension

The general assumptions in this section are the same as those in the preceding section.

14.9 Lemma
For every $\varepsilon > 0$,

$$\inf\{P(B(z,\varepsilon)) \colon z \in A\} > 0. \tag{14.13}$$

Proof

Set $t = (\frac{\varepsilon}{\operatorname{diam}(A)})^D$ and $\Gamma(t) = \{\sigma \in \{1, \dots, N\}^* : s_{\sigma^-}^D \geq t > s_\sigma^D\}$. Then $\Gamma(t)$ is a finite maximal antichain. Thus $a = \min\{p_\sigma : \sigma \in \Gamma(t)\} > 0$. Let $z \in A$ be arbitrary. Since $A = \bigcup_{\sigma \in \Gamma(t)} S_\sigma(A)$ there is a $\tau \in \Gamma(t)$ with $z \in S_\tau(A)$. Since S_τ is a similarity we have

$$\operatorname{diam}(S_\tau(A)) = s_\tau \operatorname{diam}(A)$$

and, hence,

$$\operatorname{diam}(S_\tau(A)) < \varepsilon.$$

Thus, we have $S_\tau(A) \subset B(z, \varepsilon)$. From (14.3) it follows that

$$\begin{aligned} P(S_\tau(A)) &= \sum_{\sigma \in \Gamma(t)} p_\sigma P \circ S_\sigma^{-1}(S_\tau(A)) \\ &\geq p_\tau P(S_\tau^{-1}(S_\tau(A))) \\ &= p_\tau. \end{aligned}$$

Combining the last two results we obtain

$$P(B(z, \varepsilon)) \geq p_\tau \geq a > 0$$

and the lemma is proved. □

14.10 Lemma

Let $(S_1, \dots, S_N)$ satisfy the strong separation condition and let $r \in [1, +\infty)$ be given. Then

$$V_{n,r}(P) = \min\left\{\sum_{i=1}^N p_i s_i^r V_{n_i,r}(P) : 1 \leq n_i, \sum_{i=1}^N n_i \leq n\right\} \tag{14.14}$$

for all but finitely many $n \in \mathbb{N}$.

Proof

"$\leq$": That $V_{n,r}(P)$ is less than or equal to the right hand side for all but finitely many n is the statement of Lemma 14.1.

"$\geq$": To show the converse inequality let $\delta = \min\{d(S_i(A), S_j(A)) : i \neq j\}$. Then we have $\delta > 0$. From Lemma 14.9 we deduce

$$\beta := \inf\left\{\left(\frac{\delta}{4}\right)^r P\left(B\left(z, \frac{\delta}{4}\right)\right) : z \in A\right\} > 0.$$

By Lemma 6.1 we know that

$$\lim_{n \to \infty} V_{n,r}(P) = 0.$$

Hence there exists an $n_0 \in \mathbb{N}$ with

$$V_{n,r}(P) < \beta$$

for all $n \geq n_0$. Let α_n be an n-optimal set of centers for P of order r. Then Lemma 13.8 implies that

$$\left(\frac{1}{2}\|d_{\alpha_n}\|_\infty\right)^r \min_{y \in A} P\Big(B\Big(y, \frac{1}{2}\|d_{\alpha_n}\|_\infty\Big)\Big) < \beta$$

for all $n \geq n_0$ and all $a \in \alpha_n$. Since the function $t \to t^r \min_{y \in A} P(B(y,t))$ is non-decreasing it follows that

$$\frac{1}{2}\|d_{\alpha_n}\|_\infty < \frac{\delta}{4}$$

i. e.

$$\|d_{\alpha_n}\|_\infty < \frac{\delta}{2}$$

for all $n \geq n_0$ and all $a \in \alpha_n$. By the definition of δ we deduce that, for $n \geq n_0$ and $i, j \in \{1, \ldots, N\}$ with $i \neq j$, the sets $\alpha_{n,i} = \{a \in \alpha_n \colon W(a|\alpha_n) \cap S_i(A) \neq \emptyset\}$ and $\alpha_{n,j} = \{a \in \alpha_n \colon W(a|\alpha_n) \cap S_j(A) \neq \emptyset\}$ are disjoint.
Using (14.3) we obtain

$$\begin{aligned} V_{n,r}(P) &= \int \min_{a \in \alpha_n} \|x - a\|^r \, dP(x) \\ &= \sum_{i=1}^N p_i \int \min_{a \in \alpha_n} \|S_i(x) - a\|^r \, dP(x) \\ &= \sum_{i=1}^N p_i \int \min_{a \in \alpha_{n,i}} \|S_i(x) - a\|^r \, dP(x) \\ &= \sum_{i=1}^N p_i s_i^r \int \min_{b \in S_i^{-1}(\alpha_{n,i})} \|x - b\|^r \, dP(x) \\ &\geq \sum_{i=1}^N p_i s_i^r V_{n_i,r}(P), \end{aligned}$$

where $n_i = |\alpha_{n,i}| \geq 1$.
Since

$$\sum_{i=1}^N n_i = |\alpha_n| = n$$

we deduce

$$V_{n,r}(P) \geq \min\Big\{\sum_{i=1}^N p_i s_i^r V_{n_i,r}(P) \colon 1 \leq n_i, \ \sum_{i=1}^N n_i \leq n\Big\}.$$

□

14.11 Proposition

Let $(S_1 \dots, S_N)$ satisfy the strong separation condition and let $r \in [1, +\infty)$ be given. Moreover, let κ_r satisfy (14.10). Then

$$\liminf_{n\to\infty} n e_{n,r}^{\kappa_r} > 0,$$

in particular, the lower quantization dimension $\underline{D}_r(P)$ is greater than or equal to κ_r.

Proof

Since $|A| = \infty$ we have $V_{n,r}(P) > 0$ for all $n \in \mathbb{N}$. Let $n_0 \in \mathbb{N}$ be such that (14.14) holds for all $n \geq n_0$. Choose $c > 0$ with

$$V_{n,r}(P) \geq cn^{-\frac{r}{\kappa_r}}$$

for all $n < n_0$.
We will show by induction on n that

$$V_{n,r}(P) \geq cn^{-\frac{r}{\kappa_r}} \tag{14.15}$$

for all $n \in \mathbb{N}$. Let $n \in \mathbb{N}$ be such that

$$V_{k,r}(P) \geq ck^{-\frac{r}{\kappa_r}}$$

for all $k < n$.
Using this assumption and (14.14) we obtain

$$\begin{aligned} V_{n,r}(P) &= \min\Big\{\sum_{i=1}^N p_i s_i^r V_{n_i,r}(P) \colon 1 \leq n_i,\ \sum_{i=1}^N n_i \leq n\Big\} \\ &\geq \min\Big\{\sum_{i=1}^N p_i s_i^r c n_i^{-\frac{r}{\kappa_r}} : 1 \leq n_i,\ \sum_{i=1}^N \frac{n_i}{n} \leq 1\Big\} \\ &= cn^{-\frac{r}{\kappa_r}} \min\Big\{\sum_{i=1}^N p_i s_i^r \Big(\frac{n_i}{n}\Big)^{-\frac{r}{\kappa_r}} : 1 \leq n_i,\ \sum_{i=1}^N \frac{n_i}{n} \leq 1\Big\}. \end{aligned}$$

By Lemma 6.8 we have

$$\begin{aligned} \min\Big\{\sum_{i=1}^N p_i s_i^r \Big(\frac{n_i}{n}\Big)^{-\frac{r}{\kappa_r}} : 1 \leq n_i,\ \sum_{i=1}^N \frac{n_i}{n} \leq 1\Big\} &\geq \Big(\sum_{i=1}^N (p_i s_i^r)^{\frac{\kappa_r}{r+\kappa_r}}\Big)^{\frac{r+\kappa_r}{\kappa_r}} \\ &= 1. \end{aligned}$$

Thus we get

$$V_{n,r}(P) \geq cn^{-\frac{r}{\kappa_r}}$$

and (14.15) is proved.
It follows that

$$ne_{n,r}(P)^{\kappa_r} \geq c^{\frac{\kappa_r}{r}} > 0$$

for all $n \in \mathbb{N}$, in particular

$$\liminf_{n\to\infty} ne_{n,r}(P)^{\kappa_r} > 0.$$

The remaining statement in the proposition is an immediate consequence of Proposition 11.3. □

14.12 Problem
Does the conclusion of Proposition 14.11 remain true under the weaker assumption that $(S_1, \dots, S_N)$ satisfies the open set condition?

14.13 Proposition
Let $(S_1, \dots, S_N)$ satisfy the open set condition and let D be the similarity dimension of $(S_1, \dots, S_N)$. Then

$$\liminf_{n\to\infty} ne_{n,\infty}(P)^D > 0,$$

in particular, the lower quantization dimension $\underline{D}_\infty(P)$ is greater than or equal to D.

Proof

Since $\mathrm{supp}(P) = A = \mathrm{supp}(\mathcal{H}^D_{|A})$ the statement of the proposition follows immediately from Example 12.10, Theorem 12.18 and Remark 11.2. □

14.4 The quantization dimension

The general assumptions are the same as in Section 14.2. We denote the similarity dimension of $(S_1, \dots, S_N)$ by D. In this section we will show that, for most self-similar measures, the quantization dimensions of different orders are different.

14.14 Theorem
Let $r \in [1, +\infty)$, let $\kappa_r \in (0, \infty)$ be defined by

$$\sum_{i=1}^{N} (p_i s_i^r)^{\frac{\kappa_r}{r+\kappa_r}} = 1, \tag{14.16}$$

and let $(S_1, \dots, S_N)$ satisfy the strong separation condition. Then

$$0 < \liminf_{n\to\infty} ne_{n,r}(P)^{\kappa_r} \leq \limsup_{n\to\infty} e_{n,r}(P)^{\kappa_r} < +\infty.$$

In particular, the quantization dimension $D_r(P)$ of order r exists and equals κ_r.

Proof

The result follows from Proposition 14.5 and Proposition 14.11. □

Next we will prove some auxillary results concerning the function $K\colon [1,+\infty) \to (0,+\infty)$, $r \to \kappa_r$. Define $F\colon [1,+\infty) \times (0,+\infty) \to \mathbb{R}$ by

$$F(r,t) = \sum_{i=1}^{N} (p_i s_i^r)^{\frac{t}{t+r}} - 1.$$

By definition K is the unique function on $[1,+\infty)$ with

$$F(r, K(r)) = 0$$

for all $r \in [1,\infty)$.
Since

$$\frac{\partial F}{\partial r}(r,t) = \sum_{i=1}^{N} (p_i s_i^r)^{\frac{t}{t+r}} \frac{t}{(t+r)^2}(-\log p_i + t \log s_i)$$

and

$$\frac{\partial F}{\partial t}(r,t) = \sum_{i=1}^{N} (p_i s_i^r)^{\frac{t}{t+r}} \frac{r}{(t+r)^2}(\log p_i + r \log s_i)$$

implicit differentiation yields

$$K'(r) = \frac{\kappa_r}{r} \frac{\sum_{i=1}^{N} (p_i s_i^r)^{\frac{\kappa_r}{r+\kappa_r}} (\log p_i - \kappa_r \log s_i)}{\sum_{i=1}^{N} (p_i s_i^r)^{\frac{\kappa_r}{r+\kappa_r}} (\log p_i + r \log s_i)} \tag{14.17}$$

14.15 Lemma
If there exists an $r_0 \geq 1$ with $K'(r_0) = 0$ then $p_i = s_i^D$ for $i = 1, \ldots, N$ and $K_r = D$ for all $r \in [1,+\infty)$.

Proof

Set $q_i = (p_i s_i^{r_0})^{\frac{\kappa_{r_0}}{r_0+\kappa_{r_0}}}$. Since $K'(r_0) = 0$ we derive from (14.17) and $\kappa_{r_0} > 0$ that

$$\sum_{i=1}^{N} q_i(\log p_i - \kappa_{r_0} \log s_i) = 0. \tag{14.18}$$

By the definition of q_i we have

$$s_i = \left(\frac{1}{p_i} q_i^{\frac{r_0+\kappa_{r_0}}{\kappa_{r_0}}}\right)^{\frac{1}{r_0}} \tag{14.19}$$

and hence

$$\begin{aligned}
\kappa_{r_0} \log s_i &= \frac{\kappa_{r_0}}{r_0}\left(-\log p_i + \frac{r_0 + \kappa_{r_0}}{\kappa_{r_0}} \log q_i\right) \\
&= -\frac{\kappa_{r_0}}{r_0} \log p_i + \frac{r_0 + \kappa_{r_0}}{r_0} \log q_i.
\end{aligned}$$

Substituting this value into (14.18) yields

$$\sum_{i=1}^{N} q_i\left(\frac{r_0+\kappa_{r_0}}{r_0}\log p_i - \frac{r_0+\kappa_{r_0}}{r_0}\log q_i\right) = 0.$$

Since $\frac{r_0+\kappa_{r_0}}{r_0} > 0$ this implies

$$\sum_{i=1}^{N} q_i \log\frac{p_i}{q_i} = 0.$$

Since the logarithm is a strictly concave function this implies

$$p_i = q_i$$

for $i = 1, \ldots, N$. Hence (14.19) yields

$$s_i = p_i^{\frac{1}{\kappa_{r_0}}}$$

i. e.

$$p_i = s_i^{\kappa_{r_0}}$$

Since $\sum_{i=1}^{N} p_i = 1$ the definition of the similartiy dimension of $(S_1, \ldots, S_N)$ yields

$$\kappa_{r_0} = D$$

and $p_i = s_i^D$ for $i = 1, \ldots, N$.
Using this identitiy in (14.16) we obtain

$$\sum_{i=1}^{N}\left(s_i^{D+r}\right)^{\frac{\kappa_r}{r+\kappa_r}} = 1.$$

This implies

$$D = \kappa_r \frac{D+r}{\kappa_r + r}$$

and, hence

$$\kappa_r = D$$

for all $r \in [1, +\infty)$. □

14.16 Lemma
If $(p_1, \ldots, p_N) = (s_1^D, \ldots, s_N^D)$ then $\kappa_r = D$ for all $r \in [1, +\infty)$.
If $(p_1, \ldots, p_n) \neq (s_1^D, \ldots, s_N^D)$ then $K\colon (0, +\infty) \to \mathbb{R}$, $r \to \kappa_r$ is strictly increasing with

$$\lim_{r\to\infty} \kappa_r = D.$$

Proof

If $(p_1, \ldots, p_N) = (s_1^D, \ldots, s_N^D)$ then the last part of the proof of Lemma 14.15 shows that $\kappa_r = D$ for all $r \in [1, +\infty)$.
If $(p_1, \ldots, p_N) \neq (s_1^D, \ldots, s_N^D)$ then it follows from Lemma 14.15 that $K'(r) \neq 0$ for all $r \in [1, +\infty)$. Since, as a consequence of the definition of κ_r, the function K is increasing this implies that K is strictly increasing. In particular $\kappa_\infty = \lim_{r \to \infty} \kappa_r$ exists in $[0, +\infty]$. Since

$$1 = \sum_{i=1}^{N} (p_i s_i^r)^{\frac{\kappa_r}{r+\kappa_r}} \leq \sum_{i=1}^{N} s_i^{\frac{r\kappa_r}{r+\kappa_r}}$$

we obtain

$$\frac{r\kappa_r}{r+\kappa_r} \leq D,$$

hence

$$\kappa_r \leq \frac{Dr}{r-D}$$

for $r > D$. Thus we deduce

$$\kappa_\infty \leq D.$$

Since

$$\begin{aligned} 1 &= \lim_{r \to \infty} \sum_{i=1}^{N} (p_i s_i^r)^{\frac{\kappa_r}{r+\kappa_r}} = \lim_{r \to \infty} \sum_{i=1}^{N} p_i^{\frac{\kappa_r}{r+\kappa_r}} s_i^{\frac{rK_r}{r+\kappa_r}} \\ &= \sum_{i=1}^{N} s_i^{\kappa_\infty} \end{aligned}$$

we deduce

$$\kappa_\infty = D.$$

□

14.17 Theorem
Let $q, r \in [1, +\infty]$ and let $(S_1, \ldots, S_N)$ satisfy the strong separation condition.

(i) If $(p_1, \ldots, p_n) = (s_1^D, \ldots, s_n^D)$ then the quantization dimension $D_r(P)$ of order r exists and equals D.

(ii) If $(p_1, \ldots, p_N) \neq (s_1^D, \ldots, s_N^D)$ then the quantization dimension $D_r(P)$ exists and

$$D_r(P) = \begin{cases} D, & r = +\infty \\ \kappa_r, & r < +\infty. \end{cases}$$

Moreover,

$$q < r \Rightarrow D_q(P) < D_r(P)$$

and

$$\lim_{r \to \infty} D_r(P) = D.$$

Proof

That $D_\infty(P)$ exist and equals D follows from Proposition 14.7 and Proposition 14.13. The remaining statements follow from Theorem 14.14 and Lemma 14.16. □

14.18 Problem
Does Theorem 14.17 hold under the weaker assumption that $(S_1, \ldots, S_N)$ satisfies the open set condition?

14.19 Remark
It is shown by Kawabata and Dembo (1994, Theorem 4.1), that under the assumptions of Theorem 14.17 the rate distortion dimension of P equals

$$\frac{\sum_{i=1}^{N} p_i \log p_i}{\sum_{i=1}^{N} p_i \log s_i}$$

and, therefore, equals the Hausdorff dimension $\dim_H(P)$ of P by Cawley and Mauldin (1992, Theorem 2.1).

14.5 The quantization coefficient

In the preceding sections we have shown that, for many self-similar probabilities P, the inequality

$$0 < \liminf_{n\to\infty} ne_{n,r}(P)^{D_r} \leq \limsup_{n\to\infty} ne_{n,r}(P)^{D_r} < +\infty$$

holds for $r \in [1, \infty]$ and the quantization dimension D_r of order r for P. It is, therefore, natural to investigate the problem under what conditions the above sequence has a finite and positive limit. Taking $r < \infty$ and generalizing Theorem 6.2 and (6.4) the $\frac{r}{D_r}$-th power of this limit, if it exists, is called the r-th quantization coefficient $Q_r(P)$. For $r = \infty$, $\lim_{n\to\infty} n^{\frac{1}{D_r}} e_{n,\infty}$, if it exists, is called the covering coefficient or quantization coefficient of order ∞ and does only depend on supp(P) ((10.10) and Theorem 10.7). Little seems to be known about the above problem. We will first state a positive result concerning the quantization coefficient of order ∞. To this end we need the following definition.

14.20 Definition
An N-tuple $(s_1, \ldots, s_N)$ of real numbers is called **arithmetic** if there is a positive $s \in \mathbb{R}$ with $s_1, \ldots, s_N \in s\mathbb{Z} := \{sn \colon n \in \mathbb{Z}\}$.

14.21 Theorem
Let $(S_1, \ldots, S_N)$ be an N-tuple of contracting similarity transformations of $\mathbb{R}^d$ statisfying the open set condition and let the corresponding N-tuple $(s_1, \ldots, s_N)$ of contraction numbers be such that $(\log s_1, \ldots, \log s_N)$ is not arithmetic. Let A be the

attractor of $(S_1, \ldots, S_N)$ and D its similarity dimension. Then $(ne_{n,\infty}(A)^D)_{n\in\mathbb{N}}$ has a finite and positive limit, hence the quantization coefficient $Q_\infty(A)$ of A exists in $(0, +\infty)$.

Proof

With an argument similar to that given in Remark 10.9 one can show that $\lim_{n\to\infty} ne_{n,\infty}(A)^D$ exists if and only $\lim_{\varepsilon\to 0} N(\varepsilon)\varepsilon^D$ exists, where $N(\varepsilon)$ is the minimal number of balls of radius $\varepsilon > 0$ that cover A. If one of the limits exists then so does the other and they agree. Due to a result of Lalley (1988) combined with a result of Schief (1994), $\lim_{\varepsilon\to 0} N(\varepsilon)\varepsilon^D$ exists in $(0, +\infty)$ under the assumptions of the theorem (see also Falconer, 1997, p. 123, Proposition 7.4). □

The following proposition shows that the quantization coefficient $Q_\infty(A)$ need not exist if the assumption is dropped that $(\log s_1, \ldots, \log s_N)$ is not arithmetic.

14.22 Proposition

Let $N \geq 2$, $(S_1, \ldots, S_N)$, A, and D be as above but assume that $(S_1, \ldots, S_N)$ satisfies the strong separation condition and that all S_i have the same contraction number s. Then

$$0 < \liminf_{n\to\infty} ne_{n,\infty}(A)^D < \limsup_{n\to\infty} ne_{n,\infty}(A)^D < +\infty.$$

Hence the quantization coefficient $Q_\infty(A)$ does not exist.

Proof

According to Proposition 14.7 and 14.13 we know that

$$0 < \liminf_{n\to\infty} ne_{n,\infty}(A)^D \leq \limsup_{n\to\infty} ne_{n,\infty}(A)^D < \infty.$$

Therefore, it remains to show that $(ne_{n,\infty}(A)^D)_{n\in\mathbb{N}}$ does not converge. Let $n_0 \in \mathbb{N}$ satisfy $n_0 \geq N$ and

$$e_{n_0,\infty}(A) < \frac{1}{2}\min\{d(S_i(A), S_j(A)) : i \neq j\}.$$

Using Lemma 10.2(b) it follows immediately that, for $n \geq n_0$,

$$(14.20) \quad e_{n,\infty}(A) = e_{n,\infty}\Big(\bigcup_{i=1}^N S_i(A)\Big) = \min\Big\{\max_{1\leq i\leq N} se_{n_i,\infty}(A) : 1 \leq n_i, \sum_{i=1}^n n_i \leq n\Big\}$$

We claim that

$$(14.21) \quad e_{n,\infty}(A) = se_{[\frac{n}{N}],\infty}(A)$$

for all $n \geq n_0$, where $[\frac{n}{N}]$ denotes the greatest integer less than or equal to $\frac{n}{N}$. Setting $n_i = [\frac{n}{N}]$ it follows from (14.20) that

$$e_{n,\infty}(A) \leq se_{[\frac{n}{N}],\infty}(A).$$

Now let $n_1, \ldots, n_N \in \mathbb{N}$ satisfy $1 \leq n_i$, $\sum_{i=1}^{N} n_i \leq n$, and

$$e_{n,\infty}(A) = \max_{1 \leq i \leq N} s e_{n_i,\infty}(A).$$

Without loss of generality we assume $n_1 \leq n_2 \leq \ldots \leq n_N$ so that

$$e_{n,\infty}(A) = s e_{n_1,\infty}(A)$$

and $n_1 \leq [\frac{n}{N}]$. Thus we deduce

$$e_{n,\infty}(A) = s e_{n_1,\infty}(A) \geq s e_{[\frac{n}{N}],\infty}(A)$$

and our claim is proved.
Since $N \geq 2$ the equality (14.21) implies

$$e_{2n_0+1,\infty}(A) = s e_{n_0,\infty}(A) = e_{2n_0,\infty}(A) \tag{14.22}$$

and, therefore,

$$e_{N^k(2n_0+1),\infty}(A) = e_{N^k(2n_0),\infty}(A) \tag{14.23}$$

for all $k \in \mathbb{N}$. Now assume that $\left(n e_{n,\infty}(A)^D\right)_{n \in \mathbb{N}}$ converges to some constant $c \in (0, \infty)$. Using (14.23) this implies

$$\begin{aligned} c &= \lim_{k \to \infty} (N^k(2n_0+1)) e_{N^k(2n_0+1),\infty}(A)^D \\ &= \lim_{k \to \infty} (N^k(2n_0+1)) e_{N^k(2n_0),\infty}(A)^D \\ &= \lim_{k \to \infty} \left(\frac{2n_0+1}{2n_0}\right) (N^k(2n_0)) e_{N^k(2n_0),\infty}(A)^D \\ &= \left(\frac{2n_0+1}{2n_0}\right) c, \end{aligned}$$

which yields a contradiction and finishes the proof of the proposition. □

14.23 Remark

It remains an open problem to characterize those self-similar sets A for which the quantization coefficient $Q_\infty(A)$ exists by a natural condition on the generating N-tuple $(S_1, \ldots, S_N)$. For $1 \leq r < \infty$ and general self-similar probabilities P almost nothing is known about the existence of the quantization coefficients $Q_r(P)$. The only self-similar probability P for which the existence of all quantization coefficients $Q_r(P)$, $1 \leq r \leq \infty$ is known seems to be the restriction of Lebesgue measure to the unit cube in $\mathbb{R}^d$. The classical Cantor distribution P on $\mathbb{R}$ has no quantization coefficient $Q_2(P)$ (cf. Graf and Luschgy, 1997).

Below we will summarize the known results for the Cantor distribution.
Let $S_1, S_2 \colon \mathbb{R} \to \mathbb{R}$ be defined by $S_1 x = \frac{1}{3}x$ and $S_2 x = \frac{1}{3}x + \frac{2}{3}$. Then S_1 and S_2 are similarity transformations with contraction number $\frac{1}{3}$. The attractor of the pair

(S_1, S_2) is the classical **Cantor set** $C \subset [0,1]$. The similarity dimension of (S_1, S_2) equals $D = \frac{\log 2}{\log 3}$. Let P be the self-similar probability corresponding to $(S_1, S_2, \frac{1}{2}, \frac{1}{2})$ (see (14.3)). According to (14.4) P is the normalized D-dimensional Hausdorff measure on C. This distribution is called the (classical) **Cantor distribution**. Since (S_1, S_2) satisfies the strong separation condition and since $(s_1^D, s_2^D) = (\frac{1}{2}, \frac{1}{2})$ we know from Theorem 14.17 that D is the quantization dimension of P of order r for all $r \in [1, +\infty]$. In the following theorem we will describe all optimal sets of n-centers, the quantization errors $V_{n,r}(P)$, and all limits points of the sequence $(n^{2/D} V_{n,r}(P))_{n \in \mathbb{N}}$ for $r = 2$. In particular we show that the quantization coefficient $Q_2(P)$ does not exist.

To do this we need some more notation. For $\sigma \in \{1,2\}^*$ let $a_\sigma = S_\sigma(\frac{1}{2})$. For $n \geq 1$ let $l(n) = [\log_2 n]$. For $I \subset \{1,2\}^{l(n)}$ with $|I| = n - 2^{l(n)}$ let

$$\alpha_n(I) = \left\{ a_\sigma : \sigma \in \{1,2\}^{l(n)} \setminus I \right\} \cup \bigcup_{\sigma \in I} \{a_{\sigma 1}, a_{\sigma 2}\}$$

Define $f \colon [1,2] \to \mathbb{R}$ by

$$f(x) = \frac{1}{72} x^{\frac{2}{D}} (17 - 8x).$$

14.24 Theorem

Let P be the Cantor distribution and let D, $l(n)$, $\alpha_n(I)$, and f be defined as above.

(a) For every natural number $n \geq 1$ the following conditions are equivalent

(i) α is an n-optimal set of centers of order 2 for P

(ii) There exits an $I \subset \{1,2\}^{l(n)}$ with $\alpha = \alpha_n(I)$

(b) For every natural number $n \geq 1$,

$$V_{n,2}(P) = \frac{1}{18^{l(n)}} \cdot \frac{1}{8} \left(2^{l(n)+1} - n + \frac{1}{9}(n - 2^{l(n)}) \right).$$

(c) The set of all accumulation points of $(n^{\frac{2}{D}} V_{n,2}(P))_{n \in \mathbb{N}}$ is the intervall

$$\left[\frac{1}{8}, f\left(\frac{17}{8 + 4D} \right) \right] = [0.125, 0.2589\ldots].$$

(Notice that $\frac{1}{8} = V_2(P)$.)
In particular P has no second quantization coefficient.

The proof is given in Graf and Luschgy (1997) and will be omitted here.

Notes

The definition of self-similarity as used in this section was introduced by Hutchinson (1981). His paper also contains the basic results about self-similar sets and measures. Other references concerned with this subject are the books of Barnsley (1988), Falconer (1990, 1997), and Mattila (1995). The book of Barnsley (1988) describes many interesting examples of self-similar sets and measures. The idea of studying the quantization of self-similar probabilities goes back to Zador (1982). But his results are not formulated in a rigorous way. Since then nobody seems to have dealt with the problem. Thus, all the quantization results in this Section 14 seem to be new.

Appendix

Univariate distributions

The following univariate distributions served as examples. Recall that the r-th (absolute) moment about the center of a real random variable X is given by

$$V_r(X) = \inf_{a \in \mathbb{R}} E|X - a|^r.$$

Normal distribution $N(0, \sigma^2)$

The normal distribution is strongly unimodal. If X is $N(0, \sigma^2)$-distributed, then

$$V_r(X) = E|X|^r = \sqrt{\frac{2^r \sigma^{2r}}{\pi}} \Gamma\Big(\frac{r+1}{2}\Big), \; r \geq 1.$$

In particular

$$V_1(X) = \sigma\sqrt{2/\pi}.$$

Logistic distribution $L(a)$

The density (with respect to λ) is given by

$$h(x) = \frac{e^{x/a}}{a(1 + e^{x/a})^2}, \; x \in \mathbb{R},$$

where $a > 0$ is a scale parameter. The logistic distribution is symmetric about the origin and strongly unimodal. The distribution function takes the form

$$F(x) = \frac{1}{1 + e^{-x/a}} = \frac{e^{x/a}}{1 + e^{x/a}}.$$

Suppose that X has distribution $L(a)$. Then

$$\begin{aligned} V_r(X) = E|X|^r &= 2a^r \Gamma(r+1) \sum_{j=1}^{\infty} (-1)^{j-1} j^{-r}, \; r \geq 1 \\ &= 2a^r \Gamma(r+1)(1 - 2^{-(r-1)})\zeta(r), \; r > 1, \end{aligned}$$

where ζ denotes the Riemann zeta function. In particular

$$V_1(X) = 2a \log 2, \; V_2(X) = \frac{a^2\pi^2}{3}.$$

Generalized Logistic distribution $GL(a, b)$

The density is defined by

$$h(x) = \frac{\Gamma(2/b)e^{x/ab}}{a\Gamma(1/b)^2(1 + e^{x/a})^{2/b}}, \; x \in \mathbb{R},$$

where $a > 0$, $b > 0$. We have $GL(a, 1) = L(a)$.

Double Exponential distribution $DE(a)$

The density is given by

$$h(x) = \frac{1}{2a} e^{-|x|/a}, \; x \in \mathbb{R},$$

where $a > 0$ is a scale parameter. The double exponential distribution is strongly unimodal. The distribution function takes the form

$$F(x) = \begin{cases} \frac{1}{2}e^{x/a}, & x \leq 0 \\ 1 - \frac{1}{2}e^{-x/a}, & x > 0. \end{cases}$$

If X is $DE(a)$-distributed, then

$$V_r(X) = E|X|^r = a^r\Gamma(1 + r), \; r \geq 1.$$

In particular

$$V_1(X) = a \, , \; V_2(X) = 2a^2.$$

Double Gamma distribution $D\Gamma(a, b)$

The density is given by

$$h(x) = \frac{1}{2a^b\Gamma(b)} |x|^{b-1} e^{-|x|/a}, \; x \in \mathbb{R},$$

where $a > 0$ is a scale parameter and $b > 0$ is a shape parameter. We have $D\Gamma(a, 1) = DE(a)$. If X is $D\Gamma(a, b)$-distributed, then

$$V_r(X) = E|X|^r = \frac{a^r\Gamma(b + r)}{\Gamma(b)}, \; r \geq 1.$$

In particular

$$V_1(X) = ab, \; V_2(X) = a^2b(b + 1).$$

Hyper-exponential distribution $HE(a,b)$

The density is given by

$$h(x)=\frac{b}{2a\Gamma(1/b)}\exp\left(-\left|\frac{x}{a}\right|^{b}\right),\ x\in\mathbb{R},$$

where $a>0$ is a scale parameter, $b>0$. The hyper-exponential distribution is strongly unimodal if $b\geq 1$. We have $HE(a,1)=DE(a)$ and $HE(a,2)=N(0,a^2/2)$. Let X be $HE(a,b)$-distributed. Then

$$V_r(X)=E|X|^r=\frac{a^r\Gamma(\frac{r+1}{b})}{\Gamma(\frac{1}{b})},\ r\geq 1.$$

Uniform distribution $U([a,b])$

The density is given by

$$h(x)=\frac{1}{b-a}1_{[a,b]}(x),$$

where $a,b\in\mathbb{R}$, $a<b$. The uniform distribution is symmetric about $(a+b)/2$ and strongly unimodal. Let X be $U([a,b])$-distributed.Then

$$\operatorname{Med}(X)=\{(a+b)/2\},\ EX=(a+b)/2,$$
$$V_r(X)=E\left|X-\frac{a+b}{2}\right|^r=\frac{(b-a)^r}{(1+r)2^r},\ r\geq 1.$$

Triangular distribution $T(a,b;c)$

The density is given by

$$h(x)=\frac{2(x-a)}{(c-a)(b-a)}1_{[a,c]}(x)+\frac{2(b-x)}{(b-c)(b-a)}1_{[c,b]}(x),$$

where $a<c<b$. Consider the case $c=(a+b)/2$. Then the triangular distribution is symmetric about $(a+b)/2$ and

$$h(x)=\frac{4}{(b-a)^2}\big((x-a)1_{[a,(a+b)/2]}(x)+(b-x)1_{((a+b/2,b]}(x)\big).$$

If X is $T(a,b;\frac{a+b}{2})$-distributed, then

$$\begin{aligned}V_r(X)&=E\left|X-\frac{a+b}{2}\right|^r\\&=\frac{(b-a)^r}{(r+1)(r+2)2^{r-1}},\ r\geq 1.\end{aligned}$$

In particular

$$V_1(X)=\frac{b-a}{6},\ V_2(X)=\frac{(b-a)^2}{24}.$$

Exponential distribution $E(a)$

The density is given by

$$h(x) = \frac{1}{a} e^{-x/a} 1_{(0,\infty)}(x),$$

where $a > 0$ is a scale parameter. The exponential distribution is strongly unimodal. The distribution function takes the form

$$F(x) = \begin{cases} 1 - e^{-x/a}, & x > 0 \\ 0, & x \le 0. \end{cases}$$

If X is $E(a)$-distributed, then

$$\begin{aligned} \mathrm{Med}(X) &= \{a \log 2\}, \ EX = a, \\ V_1(X) &= E|X - a\log 2| = a \log 2, \\ V_2(X) &= \mathrm{Var}\, X = a^2. \end{aligned}$$

Weibull distribution $W(a, b)$

The density is given by

$$h(x) = \frac{b}{a^b} x^{b-1} \exp\Big(-\Big(\frac{x}{a}\Big)^b\Big) 1_{(0,\infty)}(x),$$

where $a > 0$ is a scale parameter and $b > 0$ is a shape parameter. The Weibull distribution is strongly unimodal for $b \ge 1$. The distribution function takes the form

$$F(x) = \begin{cases} 1 - \exp\Big(-\big(\frac{x}{a}\big)^b\Big), & x > 0 \\ 0, & x \le 0. \end{cases}$$

We have $W(a, 1) = E(a)$. The distribution $W(a, 2)$ is called Rayleigh distribution. Let X be $W(a, b)$-distributed. Then

$$\begin{aligned} \mathrm{Med}(X) &= \{a(\log 2)^{1/b}\}, \\ EX^r &= a^r \Gamma\big(1 + \frac{r}{b}\big), \ r \ge 0, \\ V_2(X) &= \mathrm{Var}\, X = a^2 \big(\Gamma(\frac{2}{b} + 1) - \Gamma(\frac{1}{b} + 1)^2\big). \end{aligned}$$

Let $b = 2$. Then

$$\begin{aligned} V_1(X) &= E\big|X - a\sqrt{\log 2}\big| \\ &= a\big[\sqrt{\log 2} + \sqrt{\pi}\big(\frac{3}{2} - 2\Phi(\sqrt{2 \log 2})\big)\big], \\ V_2(X) &= a^2 \big(1 - \frac{\pi}{4}\big). \end{aligned}$$

For

$$a = \big[\sqrt{\log 2} + \sqrt{\pi}\big(\frac{3}{2} - 2\Phi(\sqrt{2\log 2})\big)\big]^{-1} = 2.7027\ldots$$

one obtains $V_1(X) = 1$.

Gamma distribution $\Gamma(a,b)$

The density is given by

$$h(x) = \frac{1}{a^b\Gamma(b)} x^{b-1} e^{-x/a} 1_{(0,\infty)}(x),$$

where $a > 0$ is a scale parameter and $b > 0$ is a shape parameter. The Gamma distribution is strongly unimodal for $b \geq 1$. We have $\Gamma(a,1) = E(a)$. If X is $\Gamma(a,b)$-distributed, then

$$EX^r = \frac{a^r\Gamma(b+r)}{\Gamma(b)}, \; r \geq 0,$$
$$EX = ab, \; V_2(X) = \operatorname{Var} X = a^2 b.$$

Let $b = 2$. Then

$$\operatorname{Med}(X) = \{a \cdot 1.6783\ldots\}$$
$$V_1(X) = a \cdot 1.0517\ldots$$

and for $a = 0.9508\ldots$ one gets $V_1(X) = 1$.

Generalized Gamma distribution $G\Gamma(a,b,c)$

The density is given by

$$h(x) = \frac{c}{a^b\Gamma(\frac{b}{c})} x^{b-1} \exp\Big(-\Big(\frac{x}{a}\Big)^c\Big) 1_{(0,\infty)}(x),$$

where $a > 0$, $b > 0$, $c > 0$. We have $G\Gamma(a,b,1) = \Gamma(a,b)$.

Pareto distribution $P(a,b)$

The density is given by

$$h(x) = ba^b x^{-(b+1)} 1_{(a,\infty)}(x),$$

where $a > 0$ is a scale parameter, $b > 0$. The distribution function takes the form

$$F(x) = \begin{cases} 1 - \left(\frac{a}{x}\right)^b, & x > a \\ 0, & x \leq a. \end{cases}$$

Let X be $P(a,b)$-distributed. Then

$$\operatorname{Med}(X) = \{a2^{1/b}\},$$

$$EX^r = \frac{a^r b}{b-r}, \; b > r \geq 0,$$

$$V_1(X) = E|X - a2^{1/b}| = \frac{ab(2^{1/b} - 1)}{b-1}, \; b > 1,$$

$$V_2(X) = \operatorname{Var} X = \frac{a^2 b}{(b-2)(b-1)^2}, \; b > 2.$$

Cantor distribution

Let $C \subset [0,1]$ be the (classical) Cantor set and let $D = \frac{\log 2}{\log 3}$ be the Hausdorff dimension of C. Then the Cantor distribution P ist the normalized D-dimensional Hausdorff measure on C. Define $S_1, S_2 \colon \mathbb{R} \to \mathbb{R}$ by $S_1 x = \frac{1}{3}x$ and $S_2 x = \frac{1}{3}x + \frac{2}{3}$. Then P is the unique Borel probability on $\mathbb{R}$ with

$$P = \frac{1}{2}(P^{S_1} + P^{S_2}).$$

Let X be P-distributed. Then

$$E(X) = \frac{1}{2}, \quad V_2(X) = \operatorname{Var} X = \frac{1}{8}.$$

Bibliography

Abaya, E.F. and Wise, G.L. (1981). Some notes on optimal quantization. Proceedings of the International Conference on Communications (Denver, Colorado), 30.7.1–10.7.5. IEEE Press, New York.

Abaya, E.F. and Wise, G.L. (1984). Convergence of vector quantizers with applications to optimal quantizers. SIAM J. Appl. Math. 44, 183-189.

Abut, H., editor (1990). Vector Quantization. IEEE Press, New York.

Adams Jr., W.C. and Giesler, C.E. (1978). Quantizing characteristics for signals having Laplacian amplitude probability density function. IEEE Trans. Communications 26, 1295-1297.

Agrell, E. and Eriksson, T. (1998). Optimization of lattices for quantization. IEEE Trans. Inform. Theory 44, 1814-1828.

Anderberg, M.R. (1973). Cluster Analysis for Applications. Academic Press, San Diego.

Aurenhammer, F. (1991). Voronoi diagrams: A survey of a fundamental geometric data structure. ACM Computing Surveys 23, 345-405.

Baranovskii, E.P. (1965). Local density minima of a lattice covering of a four-dimensional Euclidean space by equal spheres. Soviet Math. Dokl. 6, 1131-1133.

Barnes, E.S. and Sloane, N.J.A. (1983). The optimal lattice quantizer in three dimensions. SIAM J. Algebraic Discrete Methods 4, 30-41.

Barnsley, M. (1988). Fractals Everywhere. Academic Press, London.

Bartlett, P.L., Linder, T., and Lugosi, G. (1998). The minimax distortion redundancy in empirical quantizer design. IEEE Trans. Inform. Theory 44, 1802-1813.

Benhenni, K. and Cambanis, S. (1996). The effect of quantization on the performance of sampling designs. Techn. Report No. 481, Center for Stoch. Processes, Univ. of North Carolina, Chapel Hill.

Bennett, W.R. (1948). Spectra of quantized signals. Bell Systems Tech. J. 27, 446-472.

Bock, H.H. (1974). Automatische Klassifikation. Vandenhoeck and Ruprecht, Göttingen.

Bock, H.H. (1992). A clustering technique for maximizing φ–divergence, noncentrality and discriminating power. Analyzing and Modeling Data, 19-36 (ed., M. Schader). Springer, Berlin.

Bollobás, B. (1972). The optimal structure of market areas. J. Economic Theory 4, 174-179.

Bollobás, B. (1973). The optimal arrangement of producers. J. London Math. Soc. 6, 605-613.

Bouton, C. and Pagès, G. (1997). About the multidimensional competitive learning vector quantization algorithm with constant gain. Ann. Appl. Probab. 7, 679-710.

Bucklew, J.A. and Cambanis, S. (1988). Estimating random integrals from noisy observations: Sampling designs and their performance. IEEE Trans. Inform. Theory 34, 111-127.

Bucklew, J.A. and Wise, G.L. (1982). Multidimensional asymptotic quantization theory with r-th power distortion measures. IEEE Trans. Inform. Theory 28, 239-247.

Calderbank, R., Forney Jr., G.D., and Moayeri, N., editors (1993). Coding and Quantization. DIMACS Vol. 14, American Mathematical Society.

Cawley, R. and Mauldin, R.D. (1992). Multifractal decomposition of Moran fractals. Adv. Math. 92, 196-236.

Cambanis, S. and Gerr, N.L. (1983). A simple class of asymptotically optimal quantizers. IEEE Trans. Inform. Theory 29, 664-676.

Cassels, J.W.S. (1971). An Introduction to the Geometry of Numbers. Second Printing. Springer, Berlin.

Chatterji, S.D. (1973). Les martingales et leurs application analytiques. Lecture Notes in Math. 307 (Ecole d' Eté de Probabilités: Processus Stochastiques), 27-135. Springer, Berlin.

Cohn, D.L. (1980). Measure Theory. Birkhäuser, Boston.

Cohort, P. (1997). Unicité d'un quantifieur localement optimal par le théorème du col. Technical Report, Labo. Probab., Univ. Paris 6.

Conway, J.H. and Sloane, N.J.A. (1993). Sphere Packings, Lattices and Groups. Second Edition. Springer, New York.

Cox, D.R. (1957). Note on grouping. J. Amer. Statist. Assoc. 52, 543-547.

Cuesta-Albertos, J.A. and Matrán, C. (1988). The strong law of large numbers for k–means and best possible nets of Banach valued random variables. Probab. Theory Related Fields 78, 523-534.

Cuesta-Albertos, J.A., Gordaliza, A., and Matrán, C. (1997). Trimmed k-means: an attempt to robustify quantizers. Ann. Statist. 25, 553-576.

Cuesta-Albertos, J.A., Gordaliza, A., and Matrán, C. (1998). Trimmed best k-nets: a robustified version of an L_∞-based clustering method. Statist. Probab. Letters 36, 401-413.

Cuesta-Albertos, J.A., Garciá-Escudero, L.A., and Gordaliza, A. (1999). Trimmed best k–nets: asymptotics and applications. Preprint.

Dalenius, T. (1950). The problem of optimum stratification. Scandinavisk Aktuarietidskrift 33, 203-213.

David, G. and Semmes, S. (1993). Analysis of and on Uniformly Rectifiable Sets. Mathematical Surveys and Monographs, Vol. 38, American Mathematical Society, Rhode Island.

David, G. and Semmes, S. (1997). Fractured Fractals and Broken Dreams. Clarendon Press, Oxford.

Dharmadhikari, S. and Joag-Dev, K. (1988). Unimodality, Convexity and Applications. Academic Press, Boston.

Diday, E. and Simon, J.C. (1976). Clustering analysis. Digital Pattern Recognition, 47–94 (ed., K.S. Fu). Springer, New York.

Elias, P. (1970). Bounds and asymptotes for the performance of multivariate quantizers. Ann. Math. Statist. 41, 1249-1259.

Eubank, R.L. (1988). Optimal grouping, spacing, stratification, and piecewise constant approximation. SIAM Review 30, 404-420.

Falconer, K.J. (1985). The Geometry of Fractal Sets. Cambridge University Press, Cambridge.

Falconer, K.J. (1990). Fractal Geometry. Wiley, Chicester.

Falconer, K.J. (1997). Techniques in Fractal Geometry. Wiley, Chicester.

Fang, K.-T. and Wang, Y. (1994). Number-theoretic Methods in Statistics. Chapman and Hall, London.

Federer, H. (1969). Geometric Measure Theory, Springer, Berlin-Heidelberg-New York.

Fejes Tóth, L. (1959). Sur la représentation d'une population infinie par un nombre fini d' éléments. Acta Math. Acad. Sci. Hung. 10, 299-304.

Fejes Tóth, L. (1972). Lagerungen in der Ebene, auf der Kugel und im Raum. Second Edition. Springer, Berlin.

Fleischer, P.E. (1964). Sufficient conditions for achieving minimum distortion in a quantizer. IEEE Int. Conv. Rec., part 1, 104-111.

Flury, B.A. (1990). Principal points. Biometrika 77, 33-41.

Forney Jr., G.D. (1993). On the duality of coding and quantization. Coding and Quantization, 1-14 (eds., R. Calderbank et al.). DIMACS Vol. 14, American Mathematical Society.

Fort, J.C. and Pagès, G. (1999). Asymptotics of optimal quantizers for some scalar distributions. Preprint.

Garciá–Escudero, L.A., Gordaliza, A., and Matrán, C. (1999). A central limit theorem for multivariate generalized trimmed k–means. Ann. Statist. 27, 1061-1079.

Gardner, W.R. and Rao, B.D. (1995). Theoretical analysis of the high rate vector quantization of *LPC* parameter. IEEE Trans. Speech Audio Processing 3, 367-381.

Garkavi, A.L. (1964). The best possible net and the best possible cross-section of a set in a normed space. Amer. Math. Soc. Translations 39, 111-132.

Gersho, A. (1979). Asymptotically optimal block quantization. IEEE Trans. Inform. Theory 25, 373-380.

Gersho, A. and Gray, R.M. (1992). Vector Quantization and Signal Compression. Kluwer, Boston.

Gilat, D. (1988). On the ratio of the expected maximum of a martingale and the L_p-norm of its last term. Israel J. Math. 63, 270-280.

Goddyn, L.A. (1990). Quantizers and the worst-case Euclidean traveling salesman problem. J. Combinatorial Theory Series B 50, 65-81.

Graf, S. and Luschgy, H. (1994a). Foundations of quantization for random vectors. Research Report No. 16, Applied Mathematics and Computer Science, University of Münster.

Graf, S. and Luschgy, H. (1994b). Consistent estimation in the quantization problem for random vectors. Trans. Twelfth Prague Conf. Inform. Theory, Stat. Decision Functions, Random Processes, 84-87.

Graf, S. and Luschgy, H. (1996). The quantization dimension of self-similar sets. Research Report No. 9, Dept. of Mathematics and Computer Science, University of Passau.

Graf, S. and Luschgy, H. (1997). The quantization of the Cantor distribution. Math. Nachrichten 183, 113-133.

Graf, S. and Luschgy, H. (1999a). Quantization for random vectors with respect to the Ky Fan metric. Submitted.

Graf, S. and Luschgy, H. (1999b). Quantization for probability measures with respect to the geometric mean error. Submitted.

Graf, S. and Luschgy, H. (1999c). Rates of convergence for the empirical quantization error. Submitted.

Gray, R.M. (1990). Source Coding Theory. Kluwer, Boston.

Gray, R.M., Neuhoff, D.L., and Shields, P.C. (1975). A generalization of Ornstein's $\bar{d}$ distance with applications to information theory. Ann. Probab. 3, 315-328.

Gray, R.M. and Davisson, L.D. (1975). Quantizer mismatch. IEEE Trans. Communications 23, 439-443.

Gray, R.M. and Karnin, E.D. (1982). Multiple local optima in vector quantizers. IEEE Trans. Inform. Theory 28, 256-261.

Gray, R.M. and Neuhoff, D.L. (1998). Quantization. IEEE Trans. Inform. Theory 44, 2325-2383.

Gruber, P. (1974). Über kennzeichnende Eigenschaften von euklidischen Räumen und Ellipsoiden I. J. Reine Angew. Math. 265, 61-83.

Gruber, P.M. and Lekkerkerker, C.G. (1987). Geometry of Numbers. Second Edition. North-Holland, Amsterdam.

Grünbaum, B. and Shephard, G.C. (1986). Tilings and Patterns. Freeman and Company, New York.

Haimovich, M. and Magnati, T.L. (1988). Extremum properties of hexagonal partitioning and the uniform distribution in euclidean location. SIAM J. Discrete Math. 1, 50-64.

Hartigan, J.A. (1978). Asymptotic distributions for clustering criteria. Ann. Statist. 6, 117-131.

Hochbaum, D. and Steele, J.M. (1982). Steinhaus's geometric location problem for random samples in the plane. Adv. Appl. Probab. 14, 56-67.

Hoffmann-Jørgensen, J. (1994). Probability with a View Toward Statistics. Vol. 1. Chapman and Hall, New York.

Hutchinson, J.E. (1981). Fractals and self-similarity. Indiana Univ. Math. J. 30, 713-747

Iyengar, S. and Solomon, H. (1983). Selecting representative points in normal populations. Recent Advances in Statistics, Papers in Honor of H. Chernoff, 579-591. Academic Press.

Jahnke, H. (1988). Clusteranalyse als Verfahren der schließenden Statistik. Vandenhoeck and Ruprecht, Göttingen.

Johnson, M.E., Moore, L.M., and Ylvisaker, D. (1990). Minimax and maximin distance designs. J. Statist. Plann. Inference 26, 131–148.

Karlin, S. (1982). Some results on optimal partitioning of variance and monotonicity with truncation level. Statistics and Probability: Essays in Honor of C. R. Rao, 375-382 (eds., G. Kallianpur et al.). North–Holland, Amsterdam.

Kawabata, T. and Dembo, A. (1994). The rate distortion dimension of sets and measures. IEEE Trans. Inform. Theory 40, 1564-1572

Kemperman, J.H.B. (1987). The median of a finite measure on a Banach space. Statistical Data Analysis based on the L_1-Norm and related Methods, 217-230 (ed., Y. Dodge). North–Holland, Amsterdam.

Kershner, R. (1939). The number of circles covering a set. Amer. J. Math. 61, 665-671.

Klein, R. (1989). Concrete and Abstract Voronoi Diagrams. Lecture Notes in Computer Science 400. Springer, New York.

Lalley, S. (1988). The packing and covering functions of some self-similar fractals. Indiana Univ. Math. J. 37, 699-709.

Lamberton, D. and Pagès, G. (1996). On the critical points of the 1-dimensional competitive learning vector quantization algorithm. Proceedings of the ESANN'96 (Bruges, Belgium), 97-101.

Li, J., Chaddha, N., and Gray, R.M. (1999). Asymptotic performance of vector quantizers with a perceptual distortion measure. IEEE Trans. Inform. Theory 45, 1082-1091.

Linder, T. (1991). On asymptotically optimal companding quantization. Problems of Control and Information Theory 20, 475-484.

Linder, T., Lugosi, G., and Zeger, K. (1994). Rates of convergence in the source coding theorem, in empirical quantizer design, and in universal lossy source coding. IEEE Trans. Inform. Theory 40, 1728-1740.

Linder, T., Zamir, R., and Zeger, K. (1999). High-resolution source coding for non-difference distortion measures: multidimensional companding. IEEE Trans. Inform. Theory 45, 548-561.

Lloyd, S.P. (1982). Least squares quantization in PCM. IEEE Trans. Inform. Theory 28, 129-137.

Lookabaugh, T.D. and Gray, R.M. (1989). High resolution quantization theory and the vector quantizer advantage. IEEE Trans. Inform. Theory 35, 1020-1033.

MacQueen, J. (1967). Some methods for classification and analysis of multivariate observations. Proc. Fifth Berkeley Symp. Math. Statist. Prob. 281-297, Univ. California Press, Berkeley.

Mann, H. (1935). Untersuchungen über Wabenzellen bei allgemeiner Minkowski Metrik. Monatsh. Math. Physik 42, 417-424.

Mattila, P. (1995). Geometry of Sets and Measures in Euclidean Spaces. Cambridge Univ. Press, Cambridge.

Max, J. (1960). Quantizing for minimum distortion. IEEE Trans. Inform. Theory 6, 7-12.

McClure, D.E. (1975). Nonlinear segmented function approximation and analysis of line patterns. Quart. Appl. Math. 33, 1-37.

McClure, D.E. (1980). Optimized grouping methods. Part 1 and part 2. Statistik Tidskrift 18, 101-110, 189-198.

McGivney, K. and Yukich, J.E. (1997). Asymptotics for geometric location problems over random samples. Preprint.

McMullen, P. (1980). Convex bodies which tile space by translation. Mathematika 27, 113-121. (Acknowledgement of priority: Mathematika 28, 191.)

Milasevic, P. and Ducharme, G.R. (1987). Uniqueness of the spatial median. Ann. Statist. 15, 1332-1333.

Møller, J. (1994). Lectures on Random Voronoi Tesselations. Lecture Notes in Statistics 87. Springer, New York.

Moran, P.A.P. (1946). Additive functions of intervals and Hausdorff measure. Proc. Cambridge Phil. Soc. 42, 15-23.

Na, S. and Neuhoff, D.L. (1995). Bennett's integral for vector quantizers. IEEE Trans. Inform. Theory 41, 886-900.

Newman, D.J. (1982). The Hexagon theorem. IEEE Trans. Inform. Theory 28, 137-139.

Niederreiter, H. (1992). Random Number Generation and Quasi-Monte Carlo Methods. CBMS-NSF Regional Conference Series in Applied Math. Vol. 63. SIAM.

Okabe, A., Boots, B. and Sugihara, K. (1992). Spatial Tesselations: Concepts and Applications of Voronoi Diagrams. Wiley, Chicester.

Pagès, G. (1997). A space quantization method for numerical integration. J. Comput. Appl. Math. 89, 1-38.

Pärna, K. (1988). On the stability of k-means clustering in metric spaces. Tartu Riikliku Ülikooli Toimetised 798, 19-36.

Pärna, K. (1990). On the existence and weak convergence of k-centres in Banach spaces. Tartu Ülikooli Toimetised 893, 17-28.

Panter, P.F. and Dite, W. (1951). Quantization distortion in pulse-count modulation with nonuniform spacing of levels. Proc. Inst. Radio Eng. 39, 44-48.

Pearlman, W.A. and Senge, G.H. (1979). Optimal quantization of the Rayleigh probability distribution. IEEE Trans. Communications 27, 101-112.

Pierce, J.N. (1970). Asymptotic quantizing error for unbounded random variables. IEEE Trans. Inform. Theory 16, 81-83.

Pisier, G. (1989). The Volume of Convex Bodies and Banach Space Geometry. Cambridge University Press, Cambridge.

Pollard, D. (1981). Strong consistency of k-means clustering. Ann. Statist. 9, 135-140.

Pollard, D. (1982a). Quantization and the method of k-means. IEEE Trans. Inform. Theory 28, 199-205.

Pollard, D. (1982b). A central limit theorem for k-means clustering. Ann. Probab. 10, 919-926.

Pötzelberger, K. and Felsenstein, K. (1994). An asymptotic result on principal points for univariate distributions. Optimization 28, 397-406.

Pötzelberger, K. (1998a). Asymptotik des Quantisierungsfehlers. Quantisierungsdimension, Verallgemeinerung des Satzes von Zador und Verteilung der Prototypen. Preprint.

Pötzelberger, K. (1998b). Asymptotik des empirischen Quantisierungsfehlers und Konsistenz des Schätzers oder Quantisierungsdimension. Preprint.

Pötzelberger, K. and Strasser, H. (1999). Clustering and quantization by MSP–partitions. Preprint.

Rachev, S.T. (1991). Probability Metrics and the Stability of Stochastic Models. Wiley, Chicester.

Rachev, S.T. and Rüschendorf, L. (1998). Mass Transportation Problems. Vol. 1 and Vol. 2. Springer, New York.

Rényi, A. (1959). On the dimension and entropy of probability distributions. Acta Math. Sci. Hung. 10, 193-215.

Rhee, W.T. and Talagrand, M. (1989a). A concentration inequality for the k-median problem. Math. Oper. Res. 14, 189-202.

Rhee, W.T. and Talagrand, M. (1989b). On the k-center problem with many centers. Oper. Res. Letters 8, 309-314.

Rogers, C.A. (1957). A note on coverings. Mathematika 4, 1-6.

Sabin, M.J. and Gray, R.M. (1986). Global convergence and empirical consistency of the generalized Lloyd algorithm. IEEE Trans. Inform. Theory 32, 148-155.

Schief, A. (1994). Separation properties for self-similar sets. Proc. Amer. Math. Soc. 122, 111-115.

Schulte, E. (1993). Tilings. Handbook of Convex Geometry, 899-932. (eds., P.M. Gruber and J.M. Wills). Elsevier Sciene Publishers.

Semadeni, Z. (1971). Banach Spaces of Continuous Functions. Polish Scientific Publishers, Warszawa.

Serinko, R.J. and Babu, G.J. (1992). Weak limit theorems for univariate k-mean clustering under a nonregular condition. J. Multivariate Anal. 41, 273-296.

Serinko, R.J. and Babu, G.J. (1995). Asymptotics of k-mean clustering under non-i.i.d. sampling. Statist. Probab. Letters 24, 57-66.

Shannon, C.E. (1959). Coding theorems for a discrete source with a fidelity criterion. IRE National Convention Record, Part 4, 142-163.

Singer, I. (1970). Best Approximation in Normed Linear Spaces by Elements of Linear Subspaces. Springer, Berlin.

Small, C.G. (1990). A survey of multidimensional medians. Int. Statist. Review 58, 263-277.

Späth, H. (1985). Cluster Dissection and Analysis. Ellis Horwood Limited, Chichester.

Stadje, W. (1995). Two asymptotic inequalities for the stochastic traveling salesman problem. Sankhyā 57, Series A, 33-40.

Steinhaus, H. (1956). Sur la division des corps matériels en parties. Bull. Acad. Polon. Sci. 4, 801-804.

Stute, W. and Zhu, L.X. (1995). Asymptotics of k–means clustering based on projection pursuit. Sankhyā 57, Series A, 462-471.

Su, Y. (1997). On the asymptotics of quantizers in two dimensions. J. Multivariate Anal. 61, 67-85.

Tarpey, T. (1994). Two principal points of symmetric, strongly unimodal distributions. Statist. Probab. Letters 20, 253-257.

Tarpey, T. (1995). Principal points and self-consistent points of symmetric multivariate distributions. J. Multivariate Anal. 53, 39-51.

Tarpey, T. (1998). Self-consistent patterns for symmetric multivariate distributions. J. Classification 15, 57-79.

Tarpey, T., Li, L., and Flury, B.D. (1995). Principal points and self-consistent points of elliptical distributions. Ann. Statist. 23, 103-112.

Tou, J.T. and Gonzales, R.C. (1974). Pattern Recognition Principles. Addison-Wesley, Reading.

Trushkin, A.V. (1984). Monotony of Lloyd's method II for log-concave density and convex error weighting function. IEEE Trans. Inform. Theory 30, 380-383.

Vajda, I. (1989). Theory of Statistical Inference and Information. Kluwer, Dordrecht.

Wagner, T.J. (1971). Convergence of the nearest neighbor rule. IEEE Trans. Inform. Theory 17, 566-571.

Webster, R. (1994). Convexity. Oxford University Press, Oxford.

Williams, G. (1967). Quantization for minimum error with particular reference to speech. Electronics Letters 3, 134-135.

Wong, M.A. (1982). Asymptotic properties of bivariate k-means clusters. Comm. Statist. Theory Methods. 11, 1155-1171.

Wong, M.A. (1984). Asymptotic properties of univariate sample k-means clusters. J. Classification 1, 255-270.

Yamada, Y., Tazaki, S., and Gray, R.M. (1980). Asymptotic performance of block quantizers with difference distortion measure. IEEE Trans. Inform. Theory 26, 6-14.

Yamamoto, W. and Shinozaki, N. (1999). On uniqueness of two principal points for univariate location mixtures. Statist. Probab. Letters 46, 33-42.

Yang, M.-S. and Yu, K.F. (1991). On a class of fuzzy c-means clustering procedures. Proceedings of the 1990 Taipei Symposium in Statistics, 635-647, (eds., M.T. Chao and P.E. Cheng). Institute of Statistical Science, Academia Sinica, Taipei.

Yukich, J.E. (1998). Probability Theory of Classical Euclidean Optimization Problems. Lecture Notes in Math. 1675. Springer, New York.

Zador, P.L. (1963). Development and evaluation of procedures for quantizing multivariate distributions. Ph. D. dissertation, Stanford Univ.

Zador, P.L. (1982). Asymptotic quantization error of continuous signals and the quantization dimension. IEEE Trans. Inform. Theory 28, 139-149.

Zemel, E. (1985). Probabilistic analysis of geometric location problems. SIAM J. Algebraic Discrete Methods 6, 189-200.

Zoppè, A. (1997). On uniqueness and symmetry of self-consistent points of univariate continuous distributions. J. Classification 14, 147-158.

Symbols

$B(a,r)$	closed ball with center a and radius r, 8
$\mathring{B}(a,r)$	open ball with center a and radius r, 165
$\mathcal{B}(\mathbb{R}^d)$	Borel sets, 20
cl	closure, 11
$C_{n,r}(P)$, $C_{n,r}(X)$	set of all n-optimal sets of centers for P (for X), 31
$C_{n,\infty}(A)$	set of all n-optimal sets of centers for A of order ∞, 137
conv α	convex hull of α, 17
$C_r(P)$, $C_r(X)$	set of all centers of P (of the random variable X) of order r, 20, 20
D_d, D_d^*	lattices, 117, 118
$DE(c)$	double exponential distribution, 67
$\det(\Lambda)$	111
$D\Gamma(a,b)$	double Gamma distribution, 99
d_H	Hausdorff metric, 57
$\operatorname{diam}(A)$	diameter of A, 24
$\underline{\dim}_B(K)$	lower box dimension of K, 158
$\overline{\dim}_B(K)$	upper box dimension of K, 158
$\dim_B(K)$	box dimension of K, 159
$\dim_H(A)$	Hausdorff dimension of A, 157
$\dim_H(P)$	Hausdorff dimension of P, 158
$\overline{\dim}_R(P)$	upper rate distortion dimension, 161
$\underline{\dim}_R(P)$	lower rate distortion dimension, 161
$\dim_R(P)$	rate distortion dimension, 162
$D_{n,r}(P)$	set of all n-optimal quantizing measures for P of order r, 59
$\underline{D}_r$, $\underline{D}_r(P)$	lower quantization dimension (of P) of order r, 155
$\overline{D}_r$, $\overline{D}_r(P)$	upper quantization dimension (of P) of order r, 155
D_r, $D_r(P)$	quantization dimension of P of order r, 155
$\underline{D}_\infty(K)$, $\overline{D}_\infty(K)$, $D_\infty(K)$	(lower, upper) quantization dimension of K of order ∞, 155

$d(x, A)$	distance from x to A, 8
$E(c)$	exponential distribution, 67
$e_{n,r}(P)$	$= V_{n,r}(P)^{1/r}$, 137
$e_{n,\infty}(A)$	n-th covering radius for A, 137
$e_{n,\infty}(P)$	138
$\mathcal{F}_n$	set of n-quantizers, 30
$GL(a, b)$	generalized logistic distribution, 99
$G\Gamma(a, b)$	generalized Gamma distribution, 99
$H(a, b)$	Leibnitz halfspace, 9
$HE(a, b)$	hyper-exponential distribution, 99
h_r	94
$\mathcal{H}^s(A)$	s-dimensional Hausdorff measure of A, 157
$\mathcal{H}^D_{\vert M}$	restriction of a (Hausdorff) measure to M, 165
$H_s(P)$	Renyi entropy of P, 133
$H(P)$, $H(X)$	differential entropy of P (of X), 133, 134
I_d	unit matrix, 54
int	interior, 9
$I(P, Q)$	average mutual information of P and Q, 161
$L(a)$	logistic distribution, 71
$\mathrm{Med}(X)$	set of medians of a real random variable X, 22
$M_{n,r}(A)$	normalized n-th quantization error for A of order r, 31
$M_{n,\infty}(A)$	138
$\mathfrak{M}_r$, $\mathfrak{M}_r(\mathbb{R}^d)$	57
$M_r(A)$	normalized r-th moment of A, 20
$M_\infty(A)$	146
$N_d(0, \Sigma)$,	d-dimensional normal distribution, 54, 106
$N(0, 1)$	normal distribution, 55
$N(\varepsilon, A)$	146
P_a	absolutely continuous part of P, 78
$P(a, b)$	Pareto distribution, 99
P^f, $P \circ f^{-1}$	image measure, 33, 162
$\mathcal{P}_n$	set of discrete probabilities with at most n points in the support, 33
P_r	94
P_s	singular part of P, 78
$Q_r(A)$	r-th quantization coefficient of A, 78, 81
$Q_r^{(L)}([0,1]^d)$	r-th lattice quantization coefficient of $[0,1]^d$, 114
$Q_r(P)$, $Q_r(X)$	r-th quantization coefficient of P (of X), 81
$Q_r^{(R)}([0,1]^d)$	r-th regular quantization coefficient of $[0,1]^d$, 110

$Q_\infty(A)$	covering coefficient of A (or quantization coefficient of order ∞), 145
$Q_\infty^{(L)}([0,1]^d)$	lattice covering coefficient of $[0,1]^d$, 148
q_σ	191
$R_{P,r}$	rate distortion function, 161
$S(a,b)$	separator of a and b, 11
$S_{n,r}(P)$, $S_{n,r}(X)$	set of all n-stationary sets of centers for P (for X) of order r, 39
$SS_{n,r}(P)$, $SS_{n,r}(X)$	set of all elements of $S_{n,r}(P)$ (of $S_{n,r}(X)$) whose Voronoi diagram is a P-tesselation, 39
S_σ	192
$\mathrm{supp}(\mu)$	topological support of a finite measure μ, 24, 165
$T(a,b;c)$	triangular distribution, 123
$U(A)$	uniform distribution on A, 20
$\mathrm{Var}\,X$	variance of a real random variable X, 22
$V_{n,r}(P)$, $V_{n,r}(X)$	n-th quantization error for P (for X) of order r, 30
$V_r(P)$, $V_r(X)$	r-th moment of P (of the random variable X), 20, 20
$W(a\|\alpha)$	Voronoi region generated by $a \in \alpha$, 8
$W(a,b)$	Weibull distribution, 73, 127
$W_0(a\|\alpha)$	open Voronoi region generated by $a \in \alpha$, 9
Δ	symmetric difference, 27
δ_x	point mass at x, 25
$\Gamma(a,b)$	Gamma distribution, 72
$\Gamma(\varepsilon)$	maximal finite antichain generated by ε, 192
λ	1-dimensional Lebesgue measure, 55
λ^d	d-dimenisonal Lebesgue measure, 13
$\mu(\cdot\|A)$	26
ρ_r	L_r-minimal metric, 33, 140
σ^-	immediate predecessor of σ, 191
$\|\sigma\|$	length of σ, 191
$\sigma_{\|m}$	restriction of σ to m, 191
$\sigma \le \tau$	σ is a predecessor of τ, 191
$\|A\|$	cardinality of A, 13
1_A	indicator function of the set A, 31
$\xrightarrow{D}$	weak convergence, 57
∂	boundary, 10
$\\|g\\|_{P,r} = \\|g\\|_r$	$L_r(P)$-norm of g, 137
$\\|h\\|_p$	$L_p(\lambda^d)$-(quasi-)norm of h, 78

$\nabla_+ f(x,y)$	one-sided directional derivative, 23
$\nabla f(x)$	gradient, 23
$[x]$	integer part of the number x, 82
$\langle x, y\rangle$	scalar product, 16
$\Vert\ \Vert$	norm on $\mathbb{R}^d$, 7

Index